Warde Fowler of Kingham

Oxford don and ornithologist

GORDON OTTEWELL

Barn Owl Books

Published by Barn Owl Books,
33 Delavale Road, Winchcombe, Cheltenham,
Gloucestershire. GL54 5YL, United Kingdom.
Tel. 01242 603464.

Printing by SS-Media Ltd.

Typesetting and production by Robert Talbot, to whom
the author is greatly indebted.

British Library Cataloguing in Publication Data.
A catalogue record for this book is available from the
British Library.
ISBN 978-0-9510586-9-5

Preface

Until being appointed head teacher at Kingham village school in 1964, I had never even heard the name of William Warde Fowler.

It took a chance encounter with a villager who, seeing me, armed with binoculars, prowling along the hedgerows, observed that I appeared to be pursuing the hobby of an illustrious former resident, to alert me to the fact that an Oxford don once lived in this quiet Oxfordshire village.

From that day on, my regard for Fowler knew no bounds. I sought for, purchased and devoured all his books on birds, interviewed elderly villagers who remembered him, and with the aid of a rare out-of-print biography by an American former pupil, began to compile my own tribute to a man who, though he had died a decade before I was born, nevertheless exerted a profound influence on my life.

This tribute, in the form of extracts from Fowler's writings and published as 'Warde Fowler's Countryside' in 1985, went some way towards ensuring that the old don's name was not entirely forgotten. However, there still remained the question of the lack of a short readable biography and this book is an attempt to fill that gap, at least as far as Fowler the villager and pioneer ornithologist is concerned.

Acknowledgments

I am greatly indebted to the Rev. Ralph Mann for his invaluable help and encouragement during the preparation of this book.

Mr Michael Lainchbury, too, was helpful and supportive throughout.

My early research at Lincoln College was made possible thanks to the kindness of the Sub-Rector, the late Rev V.H.H. Green, Dr Stephen Gill and Mrs F.M. Pinnock, librarian.

Numerous Kingham residents, and in particular members of the Phillips family, shared their recollections of Warde Fowler in the early days of the project, as did Mr Robert Aplin of Bloxham.

Sarah Edwards is to be congratulated on deciphering my much amended manuscript and thanked for offering valuable suggestions regarding presentation.

Once again I must record my gratitude to Robert Talbot, without whose expertise and infinite patience the book would have remained half forgotten in my desk drawer.

As always, Margaret my wife and Brendan, my son gave their support and encouragement.

Gordon Ottewell
Winchcombe
Gloucestershire
October 2010

Contents

1. ~ A Welsh Childhood

2. ~ From College to University

3. ~ Work and Travel

4. ~ The Gentle Don

5. ~ Quiet and Industrious Years

6. ~ 'Little Tommy'

7. ~ Of Birds and Books

8. ~ Summer Studies

9. ~ Scholar and Villager

10. ~ The Gilbert White of Kingham

11. ~ The Last Years

12. ~ A Good Man Remembered

13. ~ The Fowler Country Today

14. ~ Wildlife: A Century of Change

William Warde Fowler
from the portrait in Lincoln College hall.

CHAPTER 1

England was to me almost an unknown country, and I think it was just the accident of an early immigration to Wales that has made me always so keenly enjoy the land that is really my own.

A WELSH CHILDHOOD

Bewildered and apprehensive, the three young children huddled together for comfort as the train thundered on its way westwards. It was 1853, and the Fowler family – mother, eight-year-old John, William, two years younger, and four-year-old Alice –were leaving Staffordshire for ever, on their way to a new life in South Wales.

The memory of the train shrieking through the tunnels was to remain fresh in the mind of William almost seventy years later, when, as a lifetime of dedicated scholarship was drawing to its close, he dictated his 'Reminiscences' from his sick-bed in his beloved Kingham.

John Coke Fowler awaited his family at Merthyr Tydfil. A stipendiary magistrate and authority on colliery law, he had been newly appointed to the little industrial town and had secured possession of Gwaelod-y-Garth, a substantial house on the hillside overlooking the noisy illuminated valley. This was to be the family's home for five years, until Augusta Fowler's declining health compelled them to move southwards to Llandaff.

The presence of the nearby ironworks inevitably affected the

7

children's lives. In his old age William recorded: 'After dark we could see from our nursery windows the illumination of the valleys by the Dowlais, Penydaren, Plymouth and Cyfarthfa ironworks, and hear the din of the machinery. We were occasionally taken to see the works, and I have a picture in my mind now of the men at work running out long bars of iron with hardly any clothes on their backs. I believe I heard it said even then that their lives were short ones, that few of them lived much over forty, owing to the profuse perspiration in which they worked summer and winter.'

There can be little doubt that the liberal views and caring, compassionate nature for which he was to be noted throughout his life were fashioned and nurtured by the suffering and squalor he saw in the valley below Gwaelod-y-Garth. 'What their pay was I do not know,' he wrote years later 'but I know that their cottages were small, dreary and dirty, that they had no gardens or allotments, and that they were selling their lives very cheap to the capitalists. That the capitalist was making large sums of money out of them was a plain and obvious fact.'

Another lasting influence on William's life dating from the years on the hillside above Merthyr was his introduction to the world of natural history through the family governess, Miss Pierce. The children were taken for long walks across the moors behind the house, where they lingered searching for wild flowers on the overgrown heaps of slag from the nearby works. These flowers – celandine, coltsfoot, speedwell and stitchwort – remained firm favourites throughout his life.

On wet days, Miss Pierce entertained the children by reading to them a selection of poetry and stories. Hans Andersen's Fairy

Tales were a great favourite, and William delighted to read them aloud even as a grown man. Living as they did in an isolated neighbourhood, the Fowler children made few friends and when John eventually went off to school, William and Alice were dependent solely on one another's company. Thus began a closely-bound relationship which was to endure until Alice's death at Kingham shortly before the end of the Great War.

William Warde Fowler was born on May 16th, 1847, at Wellisford House, Langford Budville, near Wellington, Somerset. His father, John Coke Fowler, originated from Derby, where his father had been a solicitor, and his mother, Augusta Bacon, was granddaughter of John Bacon, R.A., the eminent sculptor whose works included monuments to the Lord Halifax and the Earl of Chatham in Westminster Abbey and Samuel Johnson at the entrance to the choir in St. Paul's Cathedral.

Despite basic differences in temperament, father and son enjoyed a good relationship, as is revealed in the memoir written by William after his father's death in 1900. Although himself an Oxford man, Fowler senior was of an essentially practical disposition and frequently shocked his son's sensitivities in the way he treated books: 'He read books either for amusement, or for some immediate practical purpose, and often handled them in a way that almost made my hair stand on end – turning down the pages after a fashion of his own, scribbling in them freely, and sometimes even cutting out leaves or erasing pages he disliked.'

Trout fishing was John Coke Fowler's passion, an interest which both John junior and William shared throughout the greater part of their lives. The father's enjoyment of hunting and shooting, however, never appealed to his sons; a reluctance to take life, even

9

in the interests of advancing his scientific knowledge, remained with William his whole life long and ensured his reputation as an enlightened Victorian field naturalist.

In August 1855, William and his sister were taken to Weymouth for a six week stay, with their grandfather Bacon and his mother's sister Elizabeth. Looking back, Fowler recalled: 'How pleasant to my eyes were the verdant meadows of Somersetshire after the bleak hills and smoky valleys of Glamorganshire, as the train carried us from Bristol to Taunton and Yeovil.' The railway to Weymouth had not yet been constructed and the final stretch of the journey from Yeovil was made on top of a coach: 'Not a brake nor an omnibus, but a real old-fashioned coach – painted bright red and drawn by four horses... On the way we passed the great white giant of Cerne Abbas; I remember the driver pointing it out with his whip.'

Of the many happy memories of that first seaside holiday, spent in that area of England which, next to his beloved Kingham, was to remain dearest to his heart throughout his life, Fowler cherished one particular association above all others. His aunt Elizabeth Bacon, his mother's best loved sister, made an indelible impression on the young boy's mind. Augusta Fowler was by now a sick woman, whose life was tragically drawing to its close, and although his father's second wife was to prove a dedicated and affectionate step-mother, Elizabeth Bacon remained '... the ideal of all that was beautiful, graceful and tender; and she was indeed a beautiful and intelligent woman, who was now devoting her life to the care of her old father.' Fowler went on to recall that after his mother's death, it was Elizabeth who exerted the greatest influence upon the brother and sister, and who told them about their mother, of whom they could recall but little, with feeling and affection. Years later, he was to show his friends a pencil portrait of

Elizabeth, drawn by her father in 1834, at the age of nineteen '...wearing an enormous coal-scuttle bonnet which positively seems to become her.' A life-long admirer of Jane Austen, he delighted on occasion to pass the drawing off as representing Elizabeth Bennett from *Pride and Prejudice*.

Happily for Fowler, the separation from his aunt necessitated by his return to Wales following the Weymouth holiday was to be short-lived. In 1856, Grandfather Bacon and his daughter moved to Bath, and soon afterwards first John and in the following year, William, were entered as day-boys at a private school on Bathwick Hill in that city, Elizabeth Bacon having agreed that they should stay with her during term-time.

Fowler looked back on this time with immense pleasure: 'The two years which followed were perhaps the happiest and most fruitful of my life. It was an entirely new life of freedom that my brother and I lived there, and this was true of both home and school.' The headmaster, an elderly, yet kindly and benevolent man called Kilvert, known to the boys as Gaffer, was assisted by two young Scottish ushers. Fowler suspected that the education he received might well have been considered slack. He could recall sitting no examinations, he played no games and went for no runs. Yet, despite this apparent lack of rigour, he believed that the school provided the ideal atmosphere for genuine learning to thrive: 'What I owe most to the school was the entire absence of any pressure or cramming. My brother and I ... shirked all the work we possibly could; and yet I cannot allow that we wasted our time.' Left to their own devices, the boys launched themselves into the world of learning with uninhibited zest and enthusiasm: 'We read all the books we could lay our hands on ... I recall a paper we used to produce every week or so, called the Chatterbox; the one surviving number opens with an article on the feathers of

11

butterflies' wings, written by me, with another of my brother's about some story of Greek mythology.'

The historic old city captivated them, and they spent much time exploring its streets, using the numerous church spires and towers to guide them in their wanderings. Gaining in confidence, they ventured farther afield, and were soon roaming the surrounding countryside – to Sham Castle Hill and beyond it a long ridge of down and Hampton Rocks. Butterflies, which had first begun to attract their attention amongst the slaggy mounds around Gwaelod-y-Garth, now became their passion. '...especially Skippers and Blues, our great prize being the lovely little Clifden Blue, which for many years I never saw again.'

Two remarkable phenomena occurred during 1857 which made such a profound impression on the mind of a ten-year-old boy that he was to recall them with vivid clarity over sixty years later. The first was the total eclipse of March 15th, which caused the birds to cease singing and go to roost. The second was the appearance of a great comet 'which in the autumn seemed to spread right over the heavens, and roused infinite misgivings in the minds of both adults and children.' Fowler recalled that he had been reading Harrison Ainsworth's *Lancashire Witches* at that time, 'and found Mother Demdike and Mother Chattox haunting my dreams in their nightly excursions on broomsticks.'

It was during the two years that he spent at his Aunt Elizabeth's house at Bath that Fowler's love of music, which was to play so large a part in his later life, first began to develop. Although his aunt was a fluent pianist, and gave him every encouragement, he could recall receiving no actual tuition. After his aunt had given him the notes on lines and spaces, he set to work and in a short time could read simple pieces of music: 'How it was done I cannot tell; I must have worked hard, but it was no work to me, and I have no recollection of any fatigue or labour.'

Soon afterwards, Elizabeth Bacon was to introduce her nephew to the works of Mozart, for whom Fowler was to have a special affinity, and on whom he was to become an authority. He wrote: 'When she thought I was equal to it, my Aunt proposed that we should play a duet, and selected the Andante of Mozart's E Flat Symphony. Thus my first introduction to that great composer was an attempt to play the bass of a duet in one of his most beautiful movements.' Following the removal of the family home from Merthyr to Llandaff in 1858, William was withdrawn from Gaffer Kilvert's academy without, it seems, any definite plan being made for his education before his eventual entry into public school. The boy's one consolation at saying goodbye to his aunt lay in being reunited with his sister. Alice, meanwhile, had also been revealing an aptitude for music, and had commenced studying under the organist at the nearby cathedral, an accomplished musician. William was now placed under the same master, who, after delivering a scathing criticism of the boy's self-acquired technique, proceeded to give his new pupil a thorough musical grounding.

Having established his family in the comparatively sheltered and congenial surroundings of Llandaff, John Coke Fowler determined to prepare his son personally for his forthcoming entry into public school. Like so many Victorian fathers, he appeared to evade contact with children in their infancy, preferring to exert his influence at a later stage of their development. He was, his son recorded feelingly, 'by nature somewhat austere, and apt to claim more good sense from us than children in these days would be expected to possess.' Fowler went on to recall an incident that aptly illustrated his father's approach: 'One day he was driving us out from Merthyr, when I was about eight years old. He thought he would give me a first lesson in Roman history, and told me that Rome was founded in B.C. 753. In a few minutes he asked me the date of the foundation of Rome, but as I naturally had not the faintest interest in Rome or the Romans, I had entirely forgotten it. Then he told me again, and I forgot it again, and I fear that my credit with him for intelligence was reduced to a very low ebb.'

By the time his son had reached the age of twelve, however, John Coke Fowler was more amiably disposed towards assisting with his education; in fact, Fowler recorded: 'He took his greatest pleasure in rubbing up his scholarship for the purpose.' Fowler senior was at that time engaged in writing a book on the law of collieries and, after William had completed the assignments on Latin and Greek that his father had set him, he was frequently called upon to assist with some aspect of this work.

As his father's confidence in him grew, so William was entrusted with even greater responsibility. During the summer of 1859, he was allowed to drive the family phaeton, pulled by a rather nervous bay mare, to collect his father from the Merthyr train. The magistrate was so pleased with his son's handling of the temperamental animal that he arranged for William to drive the twenty-two miles from Llandaff to Merthyr one Friday to enable them to spend a weekend fishing along the Usk valley. Elated at the trust placed in him, the twelve-year-old boy set off on the appointed day along the rough, climbing road up the Taff valley. All went well until they approached Merthyr, when a dog ran into the horse's path, causing it to take fright and bolt headlong up the dusty road, with a crowd of people in pursuit. William at last managed to control the animal, only to encounter another anxious moment shortly afterwards, when the noise and glare from the Plymouth ironworks prompted the mare to shy and threaten to overturn the phaeton. Fortunately, however, William was again able to calm the frightened horse and to complete the journey without further mishap.

Father and son were to enjoy several days fishing together during that summer before William went away to school. On another occasion, when their parents were away on holiday, William and his brother went pike fishing, taking their father's special trolling rod with a huge float attached. As an unsuccessful day drew to a close, without a single pike stirring, the boys rested the rod in the fork of a small tree while they began to pack their things to return home. Sixty years later, Fowler recalled what happened with undiminished clarity: 'Glancing at the water I suddenly saw the big float making off away from the shore. I called to my brother who was nearer to the rod than I was; he rushed down the bank, seized the rod, and tightened the line. In an instant the line cut his finger to the bone, and at the same moment far out from the bank an enormous fish leaped from the water like a salmon, then vanished beneath the waves; back came the line, and away went the float to be seen no more.'

William was to say farewell both to fishing and to family life in the following year, 1860, when he commenced his six years of boarding school education at Marlborough College.

CHAPTER 2

What I needed was to go further afield and break through the old boundaries.

FROM COLLEGE TO UNIVERSITY

William's early impressions of Marlborough were unfavourable in the extreme. Always highly sensitive to his surroundings, he found the downland scenery of North Wiltshire, which had by that time mostly undergone the transformation from sheepwalks to grain production, utterly incompatible with his inclinations. His memories, on looking back, were those of an exile banished to a foreign land: 'To me, coming from the hills and vales of Glamorganshire, the change to the open, wide-spreading downs, mostly cultivated, was very depressing, and that autumn the dreary sound of threshing machines, which I had never heard before, associated itself with what seemed to me the dreariness of the life in a way I have never forgotten.'

These sombre thoughts were obviously coloured by the thirteen-year-old boy's regret at being separated from his family, and from the comforts which both in his own home and at his Aunt Elizabeth's, had always been provided in generous measure and had naturally been taken for granted. Certainly the college food – or rather the lack of it – contributed to William's unhappiness. The boys were often so hungry, he later recalled, that: 'We found a piece of raw turnip not unpalatable when the blackberries were gone and the ripe corn harvested.' He once told a friend that he never read a newspaper account of someone being prosecuted for stealing food without remembering how he had stared longingly into a baker's window in Marlborough and been tempted to grab a handful of the mouth-watering delicacies and run. Towards the end of Fowler's time, the boys in sheer desperation staged a protest strike against the shortage of food, which probably helped towards an eventual improvement in the diet. Fowler derived no benefit from this action, however, and later maintained that the semi-starvation level of the food at Marlborough accounted for his small size and somewhat delicate physique.

Another reason why William found the adjustment to public school life difficult was the nature of his preparatory schooling. Having had virtually no experience of sports or games of any kind, he was at a distinct disadvantage as a new boy at Marlborough, which, although of a comparatively recent foundation at that time, attached considerable importance, like other schools of its kind, to competitive sports. There is no evidence to suggest that he failed in any way to make the necessary adjustment; in fact his pleasure in watching cricket probably dates from this period, but he frequently expressed his conviction in later life that sport received too great an emphasis in education generally.

Paradoxically, it was to the hard and cheerless life of these early years at Marlborough that William later attributed the love and satisfaction from the pursuit of scholarship which remained with him throughout his life. Science found no place in the public school curriculum at that time, and lacking the training and encouragement to pursue his intuitive inclinations in the field of natural history, he immersed himself in classical learning: 'I think that the want of enjoyment elsewhere forced me back on the work I had to do, and gave Caesar, Ovid and Euripides a stronger claim on my affection than they might have had if I had been happier out of doors.'

Later, his zest for learning was to gain further impetus when he entered the form of a man who was to inspire him to even greater achievement and with whom he was to form a lifelong friendship. This master was F.E. Thompson, of whom, fifty years later in his *Essays in Brief,* Fowler wrote: 'What immense enthusiasm he put into all we did together! It was that enthusiasm, expressed often rather quaintly, and always quietly, that was the secret of his power of attracting young human souls... It was this enthusiasm that made himself a happy man, and made happy all his relations with the boys, and for the best of them it remained as friendship all his life.' These words aptly summarise Fowler's own qualities as a teacher and friend a generation later.

Advancing maturity saw Fowler playing an increasingly active part in college life. The newly formed Marlborough Natural History Society provided an ideal vehicle for conveying a love of wildlife to the younger boys 'now for the first time lifted far above the old passion for bird-nesting and squirrel-killing.' His growing pleasure and ability in writing prompted him to help found *The Marlborough* magazine, of which he was the first editor. Friendships, too, developed and prospered, several of which, including that with H.A. Evans, himself to become a writer of some distinction, were to withstand the test of time.

It was now that Fowler began at last to respond to the spirit of the Marlborough region, so radically different in its structure, flora and fauna not only from his home area of Glamorgan, but also from the environs of Bath and Weymouth, the only other localities with which he was in any way familiar. 'I began to feel,' he wrote, 'the real glory of the downs, and the delight in short grass and aromatic flowers, which has made me love the south of England better than any other part of our island.' He also began at this time, to appreciate the beauty of the nearby Savernake Forest, the exploration of which provided a pleasant contrast with his rambles across the open downs.

It was the nocturnal call of a downland bird, the corncrake, that prompted Fowler to try his hand at poetry during his last years at Marlborough:

THE CORN-CRAKE

I heard the cool breeze as it came from the forest,
And kissed the broad leaves of the linden below,
And stealing their sweetness, crept on to my lattice,
And breathed all its burden upon my hot brow.

I could see the tall poplar against the dark heaven,
With a million leaves dancing, did the wind only sigh:
I could see the long ripple pass over the cornfield,
And die in the shade, as the Zephyr flew by.

Not a living thing stirred in the darkness before me,
The nightingale's head was beneath her brown wing:
If the owlet was sitting on the church tower yonder,
She stirred not, and cried not, but sat wondering.

And so still was the night, and the glimmering shadows,
That I thought some sweet sounds with its stillness might suit:
So I sat there alone and sent forth to the darkness,
Some sounds soft and low, the clear notes of my flute.

Then from under the corn and from under the ripple,
From my rose-bowered lattice not fifty yards forth,
Came a strange double craking that jarred on my fluting,

And jarred on the stillness, and woke up my wrath.

'Ah, the corn-crake!' I thought, in the still summer
twilight,
'Why of all other birds art thou only awake?'
But he craked and he craked, and I waited his stopping:
But he kept up his grating crake, crake, and crake,
crake.

So I put down my flute and I closed up my window,
And gave up the field to my foe: for I said,
'Hath this strange bird of night not more right to be
craking
Than I to be fluting just over his bed?

'Did I break on his slumber, or some golden dreaming
Of a full nest of young, a long summer of joy?
Yet to-morrow the reapers may come with the sickle,
And robbing the nest, those sweet visions destroy.

'Let him dream while he may – I am grieved if I woke
him,
To sorrow or fear, for my own sorrow's sake.'
So I lay down: but still from the cornfield I heard him
Give forth that strange grating crake, crake, and crake,
crake.

Fowler was by this time well aware that he needed to move on
to a more rigorous educational regime: 'One morning early in October 1866 with my friend Charles Crawley as a companion, I
walked over the Downs to Swindon for the last time; he went on
to Cambridge and I to Oxford. It was high time to be escaping

from the limitations of a public school, however much one loved it; I was already more than nineteen, and needed some new intellectual stimulus, some casting off of the habits of thought engendered in a place like Marlborough, far away from the real life of the world.'

Fowler matriculated to New College, Oxford in 1866, but soon after taking up his place, received word from his father that he could not afford to keep his son at Oxford unless he won a scholarship. He applied immediately for a Classical Scholarship to Lincoln College and was successful. Thus began an association with 'a small college in the very centre of the city, which at that time was in very bad repair and wanting in all the graces and elegancies of the life I had looked forward to' – an association which as undergraduate, fellow and Sub-Rector, was to absorb Fowler's energies and reflect his benign influence until well into the twentieth century.

As his *Reminiscences* reveal, Fowler launched himself into university life with single-minded determination. The physical and architectural shortcomings of Lincoln were 'entirely outweighed by the character and standing of the men among whom I found myself.' The relationships he built up with the Rector, Mark Pattison – 'The only classical scholar in Oxford at that time who really understood what was meant by learning' – with his namesake, Thomas Fowler, the Sub-Rector and with his own tutor, Henry Nettleship, were to be of a lifetime's duration – friendships based, as indeed were all those Fowler made over the years, on mutual respect and genuine affection.

Although the work was hard, the fellowship was hearty. Fowler related that Thomas Fowler's ability as a lecturer on Aristotle was equalled only by his penchant for providing excellent breakfasts: '... such as I can never forget, starting with fish and unlimited amounts of buttered toast, and went on to mutton chops, beef-steaks, and if I recollect right, a tankard of strong ale with toast in it.'

His life as an undergraduate was enriched in many ways. He enjoyed learning languages, soon read French fluently, and made a first acquaintance with German.

He discovered a liking for geology, attending lectures at the museum, and extended his knowledge of music. The lectures on ancient history however, may well have served as an early prompting towards the field of study in which he was to excel in later years.

At the end of his first year at Oxford, Fowler set off with his Marlborough friend, Herbert Evans, to walk from Oxford to Evans' home by the Severn near Chepstow, a distance of some 80 miles. After confessing that the two friends commenced 'the first of those delightfully free and happy walking expeditions which are sweetest at the age of twenty-one or so' – by catching a train to Witney, he went on to describe how, clad in college blazers and straw hats, they walked to Fairford in hot and beautiful weather. After staying overnight in the little Gloucestershire town, they inspected the celebrated stained glass in the parish church before setting off once more towards Malmesbury. After lunching at Cricklade they 'plunged into a country so desolate that we literally could not find a soul to ask our way of, or a signpost to divert us.' At length, after a friendly farmer had refreshed them with home-brewed beer, they reached Malmesbury, weary, footsore and wavering in their resolve to attain Chepstow (to appease Evans' conscience). 'Those last five miles,' Fowler recalled, 'were among the most trying that I have ever had to endure.' Looking back on his first year at Lincoln, Fowler, always an exacting judge of his own conduct and abilities, commented: 'One thing is plain from it, that I myself did not work hard enough, being too much taken up with rambling and fishing.' That he failed to get into the first class neither surprised nor alarmed him at the time: 'I was thoroughly enjoying myself, and made a resolve that in the more important trial yet to come I would not miss the mark.'

An unconnected event was soon to take place however, which although saddening for Fowler at the time, was to be instrumental in giving him independence and the means to travel. Aunt Elizabeth Bacon died and left him a third share of her estate.

CHAPTER 3

I gradually began to find out what learning meant.

WORK AND TRAVEL

Between 1867 and 1870, Fowler worked at his studies with steady determination. He once told a friend that he was fortunate in possessing the faculty to become interested in anything which he had to do; without doubt his commitment to classical studies increased as the demands of his work intensified. Despite this, he never sought prizes or distinctions. Ambition in the accepted sense was incomprehensible to him. Writing of an essay he had composed at the time for a university journal, he commented casually on its fate: 'by the way (it) was devoured by a beautiful pet greyhound on the eve of its being sent in.'

Walking continued, as ever, to provide his favourite recreation. Shotover and Boar's Hill soon became as familiar to him as the hillside walks above Merthyr – and were to remain an invaluable release from the pressures of life throughout his years at Oxford. Games exerted no appeal, and although he got on well enough with most of the sportsmen, he had nothing but scathing – and uncharacteristic – contempt for certain athletes, whom he likened to barbarians, and whose loutish behaviour was second only to 'the animal ... that wandered about the streets with nothing to do but to drink and play billiards.'

As early as 1868, Fowler's health began to give cause for concern. The accumulated effect of hours of concentrated study was proving a severe test for a none-too-strong constitution, and when his brother expressed the desire to undertake a European tour with the purpose of studying architecture, Fowler agreed to accompany him 'with no object at all except to recover health and spirits.'

This was to be the first of several such tours, made possible by the legacy left by Elizabeth Bacon, which had been divided equally between William and his brother and sister. After sight-seeing in Paris, the brothers crossed into Switzerland, visiting Geneva and Lausanne before attempting, without much success, to scale one of the lesser-known heights near Martigny, from which they descended 'very footsore and good for nothing.'

Undeterred by their first unfortunate experience of Alpine exploration, the brothers made for the Eggishorn Hotel, from which they progressed to Hospenthal, by way of the Grimsel and Furka. This was to be one of Fowler's happiest ornithological hunting grounds in later visits, when his interest in birds had eventually awakened. After driving by carriage to Lake Lucerne, the Fowlers crossed the Brünig pass and reached Meiringen, where John spent many days studying and photographing the architectural styles, while William became better acquainted with the butterflies. The return journey, made through southern Germany, provided William with the opportunity to make use of his limited German, and on his return to Oxford, filled with enthusiasm for the language, he commenced a regular series of lessons with a German teacher.

The year 1869 was to prove one of the most memorable in the whole of Fowler's life. He devoted the first six months to preparing for his degree, studying with dogged determination, yet at the same time working within the constraints of his own physical powers: 'Work must be taken steadily and quietly without hopes or fears.'

He continued to relish his Oxford walks and to enjoy the quiet fellowship of his circle of friends.

In July 1869, Fowler met two men whose friendship and influence were to remain among the most significant and valuable of his entire life. Both men were considerable older than he and neither originated from a background even remotely similar to that of the young undergraduate. One, a Swiss peasant, was to play a vital role in Fowler's introduction to the world of birds; the other, a retired Admiralty official, was instrumental in his discovery of the Cotswold village which was to be his home for almost 50 years.

It was during the last week in June, at the beginning of the long vacation, that William and John, accompanied by their friend F.H. Baynes, an experienced Alpine explorer, set off once more for Switzerland. With the previous year's happy memories of Meiringen fresh in the brothers' minds, they headed directly for that place, and asked the landlord of the hotel at which they had arranged to stay to recommend a local guide. The outcome is best told in Fowler's own words: 'He introduced a good-looking man of 45 or so, whose name was Johann Anderegg, a cousin of the famous Melchior, the prince of guides. To this man we all at once took a fancy, and at his suggestion we set out next morning for the Grimsel to explore the glaciers of the Aar.'

William, the only one of the trio to speak German, soon established a friendly relationship with their guide, and learned something of his background. Having received no formal education as a child, Anderegg had taught himself to read and write, and after a period of army service, had (like his celebrated cousin) become a guide. His prowess as a hunter and marksman had been brought to the attention of Professor Fatro of Geneva, the leading Swiss naturalist of the time, who had employed Anderegg to obtain specimens of birds and mammals for his collections and research. At this stage in their association, Anderegg's skill as a guide to the Alpine mountains and glaciers was Fowler's sole motive for hiring his services; it was not until several years had passed and the then ageing Englishman's interest in climbing had been superseded by the passion for natural history, that he was to employ Anderegg's talents as a hunter, and also as a remarkably competent field naturalist, to the full.

After a successful exploration of the Aar valley glaciers, Anderegg proposed to the Englishmen that they should tackle the Eggishorn, which they did on the following day: 'On our arrival after some adventures at the Eggishorn Hotel,' Fowler recalled, '... weary and sunburnt, we found an elderly gentleman, who eyed us with interest, next to whom I sat at the dinner which followed. We found that he was a member of the Alpine Club who had climbed Mont Blanc, Monte Rosa, and the Col du Géant and we soon struck up an acquaintance.' Next day, Fowler's face was so badly blistered by the glare of the sun and the snow that it was painful for him either to eat or speak. The gentleman, who had introduced himself as Captain Barrow, produced a jar of glycerine, which Fowler's party had omitted to pack. After travelling with his three countrymen from the Eggishorn to the Belalp and on to Zermatt, the Captain, having discovered that Fowler was at university a mere 20-odd miles from the village of Kingham, in which he lived, promised to invite him over one Sunday during the coming autumn. Fowler confessed that he dismissed the incident from his mind at the time, but 'this invitation, though I never expected to see more of him, I did actually receive, and on October 26th in that autumn found myself for the first time in the place which has long been a very dear home to me.'

Captain Barrow (a self-appointed rank, later 'promoted' to Major and eventually Colonel), although nearly 40 years Fowler's senior, was to remain a firm friend until his death, at the ripe old age of 91, almost 30 years later. His engaging eccentricities, which were to ensure him pride of place in the chapter headed 'Our Village Folks' in Fowler's *Kingham Old and New*, soon became apparent. After a good lunch, followed by stout and port wine, the Captain invited his guest to take a walk with him along the road towards Daylesford. Fowler recalled: 'During our walk he saw an old fellow with a gun to frighten the rooks away, and asked him to let him have a shot, which was duly recompensed by a shilling. In the evening, after more port and stout, he wrote up his log-book, as he called it, and the next morning completed this work by cutting out interesting extracts from *'The Times'* and pasting them in the book.' Fowler's further dealings with the Captain, especially concerning his remarkable log-book, will be referred to in greater detail in later chapters.

If 1869 was one of the most memorable years of Fowler's youth, the following year, that in which he took his degree, was to be the most crucial. Like many of his contemporaries, he found the long summer vacation, away from Oxford, provided the best conditions for concentrated study, and resolved to devote the first few weeks of the long-awaited break to that purpose. In his *Reminiscences*, he recalled his daily routine during those vital weeks: 'I used to rise at 6 or 6.30, make myself a cup of cocoa in the old nursery etna, work for an hour and a half, breakfast at nine, and then take my rod and catch a trout or two. From about half-past ten till half-past one I worked away at Herodotus or Aristotle, and then after a smoke went for walk or another fish.'

The outbreak of the Franco-Prussian war in 1870 effectively ruled out the possibility of the brothers making another journey to the Alps. Instead, thinking that a holiday in Normandy would entail no risk of involvement in the conflict, they set off late in August from Calais through Amiens and Rouen to Caen. To their dismay, while studying the castle, they were suspected of being Prussians and promptly arrested. Eventually, following much delay and general frustration, they were released and lost no time in boarding a ferry to Newhaven.

Back at Oxford, in preparation for the final degree examinations, Fowler received a letter from his father, written from Neath, to which the family had moved some years earlier: 'I am not at all expecting – not even dreaming, I may say – of a first. My notion is that you have had too many interruptions from time to time.' His son, however, despite his own self-doubts over the distractions brought on by bouts of ill-health, fulfilled his avowed aim of six years earlier; in the important trial when it came, he would not miss the mark. On 18th December, 'unless this is a dream' – he heard he had obtained a First.

CHAPTER 4

How pleasant it is to make a new friend. The time went like wildfires but we never got upon the classics.

THE GENTLE DON

Despite adhering to his resolve that work should be taken steadily and quietly, the examinations reduced Fowler to a state of near exhaustion. Early in 1871, another aspect of his indifferent health revealed itself – deafness. He had been troubled by hearing loss some years earlier, but now the problem assumed more serious proportions and despite treatment from the most eminent audiologists of the time, the trouble persisted and was to prove a handicap throughout the rest of his life.

With a first-class degree to his credit, Fowler soon received two attractive offers of employment, one of a mastership at a public school and the other of a tutorship to the family of a Scottish peer. To the surprise of his friends, however, he declined both offers. Firmly resolved to devote his life to scholarship, and knowing that there was a vacant fellowship at Lincoln College: 'I had an instinct ... that I should do better if I could catch a convenient college Fellowship, which would give me time to prosecute some aims which I had long nursed.' He was right. Soon afterwards, in April 1872, he was offered a vacant fellowship, promptly accepted, and retained the position for the 49 years until his death.

Fowler lost no time in realising his first aim. Accompanied by his friend J.A. Stewart, he set off for a month's tour of Rome, determined to study at first hand the civilisation which had intrigued him since his first elementary lessons with his father, and on which he was to become one of the foremost scholars of his day. Thomas Fowler, the Sub-Rector at Lincoln (no relation), had provided the two young men with a detailed itinerary of what they should see; reminiscing half a century later, however, Fowler declared: 'If I were going to Rome for the first time now I should abandon guide books almost entirely ... and enjoy its beauty, natural history, geology, geography and move about ... as an intelligent young human being happily enabled by Providence to see some of the most beautiful and stimulating country in the whole world.' Incredible as it seems, he never returned to Italy, finding the answers he sought in future years in the writings of the Roman scholars rather than in the streets and buildings echoing with the feet of contemporary tourists.

The summer saw Fowler alone in Vienna, indulging in his intensifying delight in music. He spent the mornings undergoing painful treatment at the hands of a celebrated Austrian ear specialist, the afternoons listening to band concerts, and the evenings at the opera. He visited the graves of Beethoven and Schubert before moving on to Salzburg, where he was permitted to play a few bars from the minuet from Don Giovanni on a piano that had once belonged to his idol, Mozart.

Back at Oxford, Fowler launched into his work as a tutor at Lincoln College with the unstinting devotion that was to be the hallmark of all he did. When not engaged on lectures, tutorials and other routine college work, his time was spent in studying all that he could find on the history of Rome in classical times. He devoted countless hours to a critical examination of the sources of all this material, filling innumerable notebooks and assimilating a wealth of knowledge on his chosen period so that he had the widest possible grasp of the subject which he had to impart to his pupils. Many years later he explained why he took so much trouble with this mastery of his subject matter: 'For me it was impossible to lecture unless I could be sure of gaining the men's attention and in order to do this I had to know the substance of my lecture and a great deal more that bore upon it, so that I could talk freely outside the limited field of my subject proper, and also keep up my interest ...year after year.'

Some measure of his success as a lecturer can be assessed from the fact that the doors of the hall at Lincoln could not be closed at times because of the sheer number of students who had crowded inside to hear him. Yet he was no showman. His strengths were his knowledge of the subject and his ability to bring it alive. Professor R.H. Coon, who studied under him, wrote: 'He never used the devices of a speaker to attract or entertain... He was quiet in manner, but the bond between him and his hearers was an intimate one... He gave them a feeling for the humanness of Rome and an understanding of Roman society as a whole.'

The cost to himself, though gladly borne, was considerable. The hours of poring over small and difficult print inflicted permanent damage to his eyesight, which, like his hearing, deteriorated steadily throughout his life. His lectures, despite their quiet mode of delivery, nevertheless proved a severe test for his physique: 'The exertion to me was so great, especially in hot or muggy weather, that I often had to go home to my rooms and change my shirt and vest just as if I had been in a football match.'

Yet, although he went to infinite pains to ensure that his lectures were stimulating and worthwhile, it was the tutorial aspect of his work that Fowler regarded as the most important and to which, in later life, he attached the greatest significance as the criterion of his achievement in his chosen sphere of influence. In common with other enlightened educators, he believed that successful teaching was based on relationships, a growing sympathy between minds. Fortunately, the existing climate at Lincoln was conducive to this approach; his friendly confiding way with his students was a logical extension of the tolerant paternalism of the Rector Mark Pattison and the Sub-Rector Thomas Fowler. But whereas Pattison chose promising students to accompany him on his walks and Thomas Fowler offered convivial hospitality at the breakfast table, Fowler endeavoured to build a lasting friendship based on mutual respect and common interest. There is ample evidence that he succeeded, and in so doing took his place among the enlightened few who demonstrated that genuine learning thrives best where warm fellowship abounds.

Fowler had been the guest of Captain Barrow at Kingham on several occasions since his first visit in the autumn of 1869. His letters to his friend, preserved in the latter's log books, together with light-hearted descriptions of walks they shared, doggerel verse and assorted observations on natural history, reveal that the young don possessed a lively sense of humour.

An example is 'The Birds to Col. Barrow':-

> Some sixty years ago, we birds
> No kindly patron knew
> And when the frosts were long and fierce,
> Our hearts dependent grew.

> At last a flagstaff raised its height,
> Majestic, to the sky,
> And farmyard into garden grew
> And all looked snug and spry.

Eight bells, before the sun was up
Rang through the frosty air,
And as our host at breakfast sat
The breadcrumbs, lo! were there.

A private soldier then he was
Or captain at the most;
(No matter which – he always stood
We're certain, at his post).

A Colonel now, he condescends
To feed us morn and night;
So when 'tis Spring for him we'll sing
With all our little might.

This document is duly signed
By Titmice, great and small,
By Robin, Nuthatch, Sparrow, with
The hearty thanks of all.

He was also adept at sketching, embellishing many of his log-book entries with pen-and-ink illustrations, of which his elderly host was particularly fond.

In 1873, after his first year as an Oxford tutor, Fowler came to the conclusion that it was no longer practical to spend his vacations at his father's home in South Wales. Remote from the libraries and amenities that he needed, he decided that the time had come to set up his own base within the vicinity of Oxford; but where? The college itself would be virtually deserted during the vacations, therefore the idea of remaining there throughout the year did not appeal to him. He recorded in his *Reminiscences* how his dilemma had a happy and surprising outcome: 'Old Mrs Lockwood, wife of the Rector of Kingham, wrote to ask me if I could find her a tenant for a little house just opposite the Rectory gate; the house was tolerable, the garden was good, the distance to the station less than a mile, the run to Oxford by train only half an hour, and the rent £30. I wrote at once and offered myself as tenant.'

The effects of this decision were both numerous and far-reaching. At the age of twenty-six and without, apparently, any inclinations towards marriage, Fowler had established himself in the countryside, yet sufficiently near to Oxford to enable him to spend weekends away, obtaining the peace and mental and physical refreshment he so badly needed. Also, such a convenient rural retreat was ideally suited to his intention to invite fellow dons, friends and students to share his hospitality. In addition, a country home provided the perfect setting for his absorbing interest in music, especially the piano, at which he had firmly resolved to become a competent performer.

Lastly, but by no means least, it was at this time that Fowler's involvement with natural history, and with birds in particular, took a significant leap forward. Hitherto, his interest, though constant from childhood, had been pursued in a desultory fashion, chiefly on journeys abroad. Now, however, first in and near Oxford, and later around this adopted village of Kingham, he discovered a deeply enriching pastime, one which offered relaxation, yet provided his enquiring mind with fresh outlets, and later prompted him to write what are still two of the most readable and absorbing bird books in the English language.

His *Reminiscences* reveal how he first embarked on his ornithological studies in Oxford during the latter half of 1873; 'I went off regularly and in almost all weathers to Christ Church Meadow, the Botanic Garden, the parks, or Mesopotamia, and got the air I needed and sometimes an interesting bird or two with it. At that time Oxford was not nearly so over-run in summer as it has been now for a long time, and I was often quite alone in these prowls... Those early morning half-hours were indeed of the greatest use to me; if I found a bird I did not know, I had to make the best of him at the time, trying to get his song in my mind, and the repetition of this for ten days or a fortnight, without the intervening aid of a book about birds, fixed the song and often the habits an appearance of the bird so firmly in my mind that to forget it was always almost impossible. Thus I learnt so much about birds without books, that though I might occasionally make blunders, I accumulated knowledge of real fact almost without knowing it... The great thing to begin with is to know your bird thoroughly well – what he says, what he does, and what he looks like.'

The kinship of these words of Fowler's with those of Gilbert White of Selborne a century earlier is strikingly unmistakable. Within a few years, in *A Year with the Birds*, Fowler was to extend the resemblance even further with his lucid and sensitive descriptions of familiar birds based on patient observation in the field, to be followed towards the end of his life by a work which set the seal on fifty years of faithful, sympathetic contemplation of man and nature around an Oxfordshire village, *Kingham Old and New*.

CHAPTER 5

*I believe indeed with all my heart that in education it is the
men and not the methods that really tell.*

"QUIET AND INDUSTRIOUS YEARS"

This was how Fowler described the period of his life from 1873
until 1881, when with his appointment as Sub-Rector at Lincoln
College and the publication of several books and learned papers,
his career was to embark upon yet another phase.

Happily established in his rented cottage at Kingham, to which
he would retire for most weekends during term-time, he set about
organising his college routine in a way best suited to his firmly-
held belief that the most important aspect of his work lay in the
meeting of minds in tutorials. He describes in *Reminiscences* how
he turned his growing deafness to advantage by inducing his
pupils to submit their work an hour or so before they themselves
were due to meet him, though it was not the recognised method.
This, he explained, provided him with the opportunity to offer a
more searching criticism of the work than if he had merely sat and
listened. 'It also enabled me to look more closely into the texture
of the pupil's work and his use of his native language.' Although
this method of working appears fairly routine today, there can be
little doubt that Fowler was in the vanguard of enlightened tutors
who believed that a working relationship could with mutual profit
be conducted on a friendly, informal level. Fowler elaborated on
his reasons for adopting this approach: 'I think ... that if he was
alone with me, as often was the case, it gave me better opportunity
of making friends with him, and discovering what manner of man
he was. If I began with a man in this way at the beginning of a

term, I knew enough of him at the end of it to ask him over to Kingham for a day or two, to work if he would, but also to be an intelligent companion to me and so far as was possible to share my occupations and pleasures. In this way some of my best friends have been made, and many of them continue my best friends to this day.'

The fact that few of these pupils won the glittering prizes at University meant nothing to Fowler: 'I used to think that they had a larger percent of human kindness in their composition than most first class men. I often have had the occasion to reflect that among my nearest and dearest old pupils there were few who achieved first classes.' This is not to say that the tutorials – or the days at Kingham – were in any way conducted in a leisurely or superficial manner.

At Lincoln, Fowler worked at his tutorials unstintingly until ten or eleven o'clock at night and would then 'drop tired into bed and sleep till morning. During those early years of my tutorship I had no special ambitions, but simply worked away from hour to hour, doing the very best I could.' The same attitude prevailed at Kingham, where although politics and economics were seldom discussed, 'we did not talk nonsense or waste time in frivolity.'

Social life at Lincoln College was virtually non-existent in the early years of Fowler's tutorship. He recalled that the atmosphere in the common room was not a happy one 'and I often sat looking at the clock and longing to get away.' Occasional social evenings at Fyfield House, the home of Mr and Mrs Acland, provided a welcome change of scene and company. Fowler, perhaps surprisingly, seems to have shone in these gatherings, where talk ranged from the gay to the serious, and readings and music were enjoyed around a bright fireside after a pleasant meal. A fellow guest remembered Fowler as 'the votary of Mozart, the lover of birds, writing essays with a felicity of style not inferior to Elia.'

But it was to his home at Kingham that Fowler turned with increasing relief and pleasure as the years passed: 'How delicious it was to run over here for a Sunday and have a really good rest, with a stroll or two in the country.' Although the village stood barely 400 feet above sea level, in the valley of the River Evenlode, some miles from the Cotswold escarpment, he expressed his unshakeable belief on numerous occasions that the air was infinitely superior to that of Oxford and was highly beneficial to his health.

His circle of friends in the village at that time was confined to Captain Barrow and the Lockwood family, who between them were responsible for introducing him to Kingham and assisting him to acquire a home there. With Barrow as his guide, he explored the lanes and footpaths for miles around Kingham, at the same time gaining an intimate knowledge of the locality, which he later used to good effect as his ornithological studies developed. Stow-on-the-Wold, the Cross Hands and the Merrymouth Inn were the most frequent destinations; here the two men could obtain refreshment before setting off on the return journey. Occasionally however, they aimed for some remote point offering no such comfort, such as Adlestrop Hill or Sarsden Pillars. The latter was a particular favourite with the ageing Captain, who in the best traditions of Alpine exploration would 'insert a visiting card in some cranny of the stonework, removing the one which he had left on his last visit.' Barrow revelled in a rigorous life and on one occasion insisted on them traversing a steep slope covered with frozen snow, using ropes and an ice axe, to the amazement of the onlooking villagers. 'The more astonished they were, the more pleased he was, and he would stop and instruct them in the art of cutting steps, ending invariably, as he ended all such conversations, with a substantial tip.' Fowler recalled that this habit of tipping was the cause of some concern in the village at that time, as it encouraged tramps from miles around to descend on Kingham to get their share of the bounty.

The young don was frequently invited to dine with one or other of his neighbours. As dinner with Captain Barrow usually came after a long and tiring walk, it was taken late and followed a familiar routine: 'Dinner lasted a long time and was followed by port wine. When at last we got back to the drawing room there was only an hour left till bed time, which was ten punctually.' Fowler usually spent that time penning a new entry in the Captain's Log Book, often in the form of an account of their afternoon's walk in the style of the Alpine Journal:

'At 4.45 pm we started for Mount Adlestrop and reached the summit without much serious difficulty and but little danger at 5.45. We stayed a short time at the hospice, where the natives were smoking foul tobacco. On our descent, we passed a camp of brigands (alias gypsies) who allowed us to pass unmolested. The view from the 'Horn' is most beautiful – a long 'sierra' stretches far in the direction of Cheltenham. The chalets of Stow standing out finely against the horizon. 6 miles.'

On another occasion, the sight of a flock of homeward-bound rooks received the full military treatment.

'We observed large corps of flying black hussars, vulgarly called rooks; the whole number must have amounted to some 1,400. They were flying in battalions towards Cornwell Park where their barracks were. The first body was very numerous, then came various smaller detachments, but all flew in the same direction and with the evident intention of getting home for tea.'

By contrast, the behaviour of a flock of starlings commanded Fowler's attention to such an extent that his log-book entry was confined to a straightforward description of what he saw:

'They were either led in all they did by some captain who was not distinguishable from the rest, or else they were all acting at once from some mysterious impulse, which forbade any individual to act on his own responsibility. They wheeled together as they flew, they lit together on the ground, and rose together from it; yet not all actually at the same moment, but in such a way as to suggest the likeness of a piece of light drapery lifted airily off the ground, waved in the air, and left to settle quickly down again.'

At Barrow's request, Fowler continued to pen accounts of his walks for some years after his host's own stamina for long rambles had passed. An example from July 1881 describes a lengthy walk along the Coln valley, accompanied by his housekeeper's two children:

'The Major desires me to report in the Log a resumé of my travels yesterday. Started at 9.40 for Andoversford (with J & S Toon). Arrived at 10.20. Set out to walk down the Coln valley, which is here very near its head. The walk is at first in open down-like country, but at the pretty village of Withington changes character – the valley narrowing and being thickly wooded on the south side. At Withington there is a fine church with a beautiful north doorway and the view of it from the valley below – where the brook begins to swarm with little trout – is most picturesque. From Withington we went on to the hamlet of Woodside, where the lane and brook run over the same ground for some 20 yards and then keeping to the left bank along two fields presently reached Cassey Compton, a fine old farmhouse. Crossing by the sheepwashing place to the right bank, we entered a part of the valley wooded on each side, with delightful views at every turn: the lane bordered on each side with a wonder of blue geraniums. Getting tired and hot by the time (12.30) we were glad to reach the Roman villa, which is exactly on the opposite side of the vale from Yanworth Common. Here we lunched, on bread and cheese and milk: after looking over the Roman remains and curiosities, left at 2.30 for Northleach by Yanworth Common, where we found the great white snail (Helix Panatia) which we kidnapped to localise at Kingham – we passed Yanworth village, dipped into a valley and rose straight on the other side and then over the high cover-ground, with views of the chalk downs of Wiltshire, to Northleach which we reached at 4 (Heat excessive in the bright, white town). Then took horse and cart from the Wheatsheaf as far as the hill over Bourton, and after waiting 1½ hours for the train, reached the Junction at 8 and Kingham at 8.20.'   W.W.F.

An evening meal with the Lockwood family too, followed its own distinctive pattern. Mrs Lockwood, the Rector's wife, was regarded as the Queen of the village, a role she upheld with gentle dignity. An invitation to dine, Fowler discovered, was virtually a command, 'though sometimes I would fain have had music at home instead.' Mrs Lockwood's after-dinner speciality were her liqueurs made from home-grown fruit; decanters of these drinks were ranged in front of her: 'The one to which I was supposed to be peculiarly addicted was damson gin.' Mrs Lockwood was the ideal consort to her husband, Samuel Davis Lockwood, whose friendship with Fowler lasted for 40 years. 'Reader, preacher, fox-hunter and helper of the distressed,' he, like Barrow, figures prominently in the chapter *Old Kingham Folks* in *Kingham Old and New.*

Such friends as these undoubtedly played their part in helping to persuade Fowler that his destinies lay forever in this little Oxfordshire village. Prompted by the questions of others, he sometimes tried to analyse what was at the root of this irresistible appeal: 'There is undoubtedly something fascinating about Kingham, as everyone who has been here will allow. Its fresh air and breezy situation have something no doubt to do with this; but there is also a certain independence and irregularity about it, making it a less common place than a village of prim cottages well looked after by a large and benevolent land-owner.' In *Kingham Old and New,* he goes on to quote with relish the reputed saying of old William Beacham, who, referring to the neighbouring village of Churchill, insisted that he would rather be hung in Kingham than die a natural death there. Whatever the reasons for Kingham's appeal, Fowler's regard for it never wavered over the years, even with the passing of those old friends whose influence brought him within its orbit.

Frequent visitors to the little house opposite the Rectory gate during Fowler's early years at Kingham were his brother John and his wife, whom he had married in 1875 and who were living at Brecon where John was an architect. This union was soon blessed with children, and Fowler was delighted to welcome them to his home. Other family visitors included his father and his sister Alice, his step-mother and his two half-sisters, Emily and Florence. His father had been consulted on the proposed move to Kingham from the outset and following the death of his gardener, William Toon, he had arranged for Mrs Toon to come to Kingham as William's housekeeper, together with her two children.

In 1877, Fowler paid another visit to the Alps, in the company of the Donkin family, with whom he had a long-standing friendship based on shared musical interest. This was to be his last mountaineering tour; his future visits to Switzerland being of an ornithological nature. With his old friend Anderegg as guide, Fowler travelled from the Eggishorn to the Grimsel, then from Meiringen to the Rhone glacier and finally across the Strahleck to Grindelwald.

In the following year, 1878, an unexpected crisis occurred which caught Fowler totally unprepared and threatened to disrupt his own life and that of the Toon family, who had severed their own connections to join him at Kingham. He had received notice from the Lockwoods that they would require his house for their parson son, who was returning to the village to assist his elderly father with his duties.

Fowler confessed to being quite perplexed by this development, but fortunately the fate that had brought him to Kingham five years before seemed bent on keeping him there. Shortly afterwards, he learned that an old farmhouse with a spacious though neglected garden and situated next to his rented house, was for sale. He promptly bought this property and enlisted the professional expertise of his brother to design a house for him to stand within the farmhouse garden. By October 1879, the new house had been built and the garden designed and partly planted, chiefly by Fowler himself, who 'set to work on it with such a good will as I have never seen before or since used in the work of my hands.'

All that now remained to be done was to restore the old farm cottage into good order to accommodate the Toon family. This accomplished, Fowler surprised his friends by acquiring a wire-haired Welsh terrier, which he called Billy, and which was to become not only something of a celebrity both at Kingham and at Oxford, but also won a wider claim to fame by featuring in *Summer Studies of Birds and Books*; *Billy, a Memoir of an Old Friend* and in *Memories of Some Oxford Pets*, collected by Mrs Wallace and published by Blackwell in 1900 to raise money to provide medical relief for men wounded in the Boer War. Fowler also contributed the preface.

At thirty-three, settled and secure, Fowler seemed destined for a quiet and unremarkable life. In fact, the years of real achievement were yet to come.

CHAPTER 6

It made me feel proud of Oxford. It made me feel the universal
fellowship of real learning, which knows no distinction of race
or language.

"LITTLE TOMMY"

William Warde Fowler was appointed to succeed Thomas
Fowler as Sub-Rector of Lincoln College in 1882 and held that
office until he resigned in 1904. During those years, he wrote three
major works on ancient classical history and four books on birds,
as well as numerous articles on birds and on a range of subjects
including Jane Austen's heroines, music in education, the role of
field voles in the Apolline worship and memoirs of his father and
of his terrier, Billy. In addition, he contributed reviews of several
books on classical history to periodicals and also published papers
for limited circulation.

During that time too, he fulfilled a leading role in the life of his
adopted village, serving as church warden and school manager,
as well as acting as counsellor, friend and benefactor to any who
sought his aid. He never married, and although he is said to have
confided to a close friend that once as an undergraduate he came
up a day late for Hilary Term in order to enjoy skating with a
young lady, and was alleged to have told a nephew that he had
once intended marriage to a lady – possibly his skating partner –
in a neighbouring village, the words 'Marriage had to be re-
nounced and forgotten' in his *Reminiscences* are the sole reference
to the subject in all his writings. Certainly, he delighted in the
company of his brother's family and his two volumes of bird stories reveal something of his natural ease in communicating with

children.

Warde Fowler was thirty-five when he succeeded his namesake Thomas as Sub-Rector at Lincoln. 'Tommy' Fowler, tall, stout, provider – and eater – of sumptuous breakfasts, had been appointed as president of Corpus. Warde Fowler described him as 'the kindest and best natured man I have almost ever known.' Thomas Fowler's relations with the Rector, Mark Pattison, had always been somewhat strained, however, and the appointment of Warde Fowler brought almost an immediate improvement in the atmosphere in the Common Room and in the college generally. This is not to say that 'Little Tommy', as Warde Fowler soon became known, found his new role particularly easy or congenial; indeed in his *Reminiscences* he alludes to several incidents in which Pattison clearly took advantage of his youthfulness and charitable disposition by leaving him to deal with problems that were not strictly within his province.

During his twenty-two years as Sub-Rector, the principal figure in the life of his college, Warde Fowler's devotion to Lincoln was total. His concern for the well-being of the undergraduates was forever foremost in his thinking, and the debt openly acknowledged to him by countless old students, many of whom reached the peaks of their chosen professions, conveys with unanimity their gratitude and affection for the deaf, slightly-built figure with the greying hair and the distinctive, almost musical way of clearing his throat.

Never one to impose his ideas in an authoritative manner, Warde Fowler eased his innovations gently into college life, where indeed they became respected and beloved traditions over the years, as did the man himself. Breakfast, for instance, was introduced into the Common Room and became the occasion for lively debate on political and other affairs of the day. One contemporary recalled: 'Fowler's talk was, like himself, full of genial insight into human beings, with an eye to the humour of things, especially when he imitated Welshmen, for he was well acquainted with Wales'.

At dinner, over which he usually presided, he took the lead in conversation and, as one listener later recalled, 'He was an excellent talker... and his deafness, just sufficient to check easy dialogue, gave him a sort of prescriptive right to have his say without interruption.' This did not mean that he monopolised conversation or that he taxed his hearers' patience; the same listener continued: 'He never bored us; his sense of fitness would not allow him to inflict on his auditors more detail than they could carry.'

And although his friendly, unassuming manner prompted him to introduce the kind of innovations into college life that made for greater informality and easier relationships, Fowler's sense of tradition remained strong and unswerving throughout his Sub-Rectorship. He stoutly resisted the introduction of electric lighting into the college, insisting on the continued use of candles in the Hall. At the same time, he refused to allow electricity into his home at Kingham. He was vehemently opposed to the felling of a plane tree adjacent to the college and campaigned vigorously in defence of the parks when a scheme was launched to encroach on the open space for extensions to the University. He also revived the traditional brewing of ale for Ascension Day at Lincoln, which had lapsed some thirty years earlier. This brew was renowned for its special infusion of ground ivy. But perhaps his sense of tradition is best illustrated by the way in which he nurtured the love of Lincoln in his undergraduates, bridging the years with his store of warm, human memories of the men he had known in his own early years. Although by temperament conservative, Fowler was, in the judgement of Dr. V.H.H. Green in *The Commonwealth of Lincoln College:* 'sensitive, tolerant and in many ways essentially liberal in attitude.' This view is borne out by Fowler's easy, conciliatory manner with people in all walks of life. For example, he recalled in *Kingham Old and New* a train journey he once took in the company of the founder of the Agricultural Labourers' Union, Joseph Arch; and that he 'talked with him the whole way, much to my profit; his personality impressed me, and he indulged in no rant, though his language was now and then a little strong.'

A friend and former pupil, Horace Mann, recorded a description of Fowler's living quarters at that time: 'His room was full of books with a black kettle on the hob, old-fashioned, much worn furniture, and a piano near the window. On the high mantelpiece stood two tins, the one of *Birds Eye* tobacco and a green one of *Three Castles* cigarettes. At the end of the fender was a dog basket, the resting place of Fowler's beloved dog.'

The one aspect of his work which he disliked, apart from the tedium of petty administration, was the matter of discipline. 'This was not much,' he wrote later, 'but it was of a tiresome nature, because a college is not a man-of-war, nor yet a public school, but something in which you must contrive to enforce rules without personal annoyance, such as is produced by penalties of the nature of fines, impositions, and so on.' The trust he bestowed on the scores of high-spirited young men who came under his care over the years was seldom misplaced. On the occasions when misdemeanours did occur, he contrived, by turning his deafness to advantage and with the aid of John Hammond, the college porter, to maintain harmony while at the same time keeping the disruption of his own routine to a minimum.

Hammond's help became virtually indispensable. He was utterly devoted to Fowler and was the only person allowed to enter the Sub-Rector's rooms in his absence when Billy, Fowler's wire-haired Welsh terrier, was in residence. Fowler openly acknowledged his debt to the porter, who 'contrived to manage the men by practical wisdom and a sense of humour. If anything illegal was about to be committed, I could trust Hammond to stop it. He would come out of his lodge with his hands in his pockets, and look at the bonfire, or whatever it was, and I was saved the necessity of coming out into the quad and trusting to my own eloquence.'

Many were the stories told of Fowler's eccentricities during his last years of Sub-Rectorship. His increasing deafness, which finally led him to resign his office, gave rise to several of these; for example, one of his pupils related how, when a newcomer to the college, he was told by Fowler what a quiet place Lincoln was, and that most of the men devoted their evenings to reading, while at the time a banjo party was taking place nearby and a rag was in progress in the quad below. Many undergraduates, hearing of the Sub-Rector's passion for birds, and interest in cricket, were said to evade censure for their misdeeds by directing conversation into one or the other of those topics. Others took advantage of Fowler's lenient attitude towards attendance at Chapel; as Lincoln's first lay-Sub-Rector and a man well known for his tolerant views, he was instinctively sympathetic on matters of personal conviction. On one occasion, on receiving an inaudible excuse from an undergraduate reported for failing to attend worship, he is said to have replied, 'Oh, of course if you have conscientious objections, I will not insist on your attendance,' thus exempting the amazed offender for the remainder of his stay.

His fellow dons enjoyed a little gentle baiting of their friend and colleague from time to time. J.A.R. Munro, later Sub-Rector himself, recalled how, when Fowler had expressed himself forcefully on a political question, 'we might enter into the game and ply him with suggestions and criticisms until we got him round to the contrary position; when we pointed out that he had demolished his own argument he would laugh and say, 'Well, I hold both opinions.'

Apart from his deafness, Fowler's health, never robust, imposed certain limitations on his way of life from the early days of his Sub-Rectorship. He recalled that his eyes began to trouble him badly during his early forties and he had no doubt as to the cause: 'I had culpably neglected them. I had for example, read through with minute care the whole of Plutarch's 48 lives in the Greek, regardless of what light I had to read by.' Later after walking with a friend the twelve miles from Kingham to the Rollright Stones and back, he suffered a form of paralysis which his doctor attributed to overwork and for which he was prescribed beefsteaks and stout, which he believed helped speed his recovery.

He was, according to his contemporaries, extremely sensitive to variations of weather throughout his life. Normally cheerful and lively, he would be irritable and moody in muggy, unsettled weather and a persistent east wind made him morose and snappish. Later, in *Kingham Old and New*, he was to draw on this acute awareness of and fascination for the phenomena of weather when describing certain memorable storms with remarkable clarity.

Although his relinquishing of the Sub-Rectorship of Lincoln College was attributed chiefly to his worsening deafness, certain other factors influenced his decision. The satisfaction he had derived from writing his early books on birds and on the history of ancient Greece and Rome intensified his desire to give more time to these interests. The death of his father, too, in 1899, played a part in shaping his future plans; his sister Alice came to share his home at Kingham and although he retained his Fellowship and continued tutoring for several years, it is clear that he considered the termination of his Sub-Rectorship as the necessary ending of one phase of life prior to embarking on another. Few men of fifty-seven have their most productive years ahead of them; Warde Fowler was one of that minority.

Before assessing the considerable achievements of these later years, however, it is necessary to return to the early years of Fowler's Sub-Rectorship, for it was then that his earliest historical and ornithological publications appeared, first fruits of what proved to be notable contributions to two widely differing, yet equally significant fields of knowledge.

CHAPTER 7

Animal life is assuredly worth study, not only in the individual as a type of his race, but in the individual as individual simply.

OF BIRDS AND BOOKS

Warde Fowler's writings on birds were published only after years of patient observation. A thorough mastery of his subject matter and a slow and deliberate assessment of his aims always preceded the committing of pen to paper. The quiet modesty of the true scholar, coupled with a natural respect for his reader remained with him throughout his writing life and lends his work an authenticity and freshness which have withstood the years with remarkable ease.

Over ten years elapsed between Fowler's earliest ornithological studies around Oxford and the publication of his first writings on the subject. These appeared in the *Oxford Magazine* during 1884, the first entitled *The Birds of Oxford City*. The author, writing at a time when bird study in and around an ancient university city was virtually unheard of, is not slow to jest about his own seeming eccentricity: 'Man in Oxford, though "featherless, beakless and human," is friendly to the birds, or at least he is indifferent to them; and that is what they enjoy. The writer is perhaps the only man in Oxford who gives them anxiety. He is apt, especially in the summer term, to look in secret places with a field-glass, and to set them a-chattering in some degree of alarm; but his habits are becoming inveterate, and they must by this time be accustomed to him.'

Although the first of the Wild Bird Protection Acts had come into force a few years earlier, persecution was still widespread and Fowler made a special plea for the birds of the Oxford parks: 'I hope that all who are friendly to our birds will do what lies in their power to induce them to remain in their present haunts, by protecting them from air-guns and catapults, and from the youthful brigands, who as the Park Keeper tells me, make inroads into the Parks at illegal hours in spite of all his watchfulness. Pleasant and exciting as it is to record the rare advent of some strange bird from fen or moorland, yet after all, our hearts incline most to the trustful little creatures who live among us the whole year round, and to the delicate travellers who come spring after spring, from distant Spain and Africa, to find a summer's shelter for themselves and their young in our abundant trees and herbage.'

Despite his failing eyesight, Fowler's interest in birds intensified after his appointment as Sub-Rector. It became his habit, whatever the state of the weather 'to steal out for twenty minutes or half an hour soon after breakfast... to let my senses exercise themselves on things outside me.' He attributed this practice to his days as a trout fisherman: 'The rod has given way to a field-glass, and the passion for killing has been displaced by a desire to see and know; a revolution which I consider has been beneficial, not only to the trout, but to myself.'

Writing, he found, taxed his eyes far less than reading and observing birds proved even less harmful. So it was, that his daily explorations of the parks, Christ Church Meadow and the Botanic Garden helped him to attain 'a knowledge of birds in their ways and their haunts which very few people at Oxford possessed at that time. My knowledge was not wide, but it was based on my own observation, and had very little to do with books. If I saw a bird that I did not know I went on observing it until I did know it; that is, if it would show itself to me I went to visit it every day with a good glass, and then whatever its name might be was matter of small moment, for I knew my bird. Its place in classification would come later.'

Fowler published more articles on bird life in the *Oxford Magazine* during 1885 and *The Zoologist* printed his *Ornithological Notes from Switzerland* based on observations made on his Alpine explorations with Anderegg. By April 1886, his bird studies around Oxford and Kingham and his Swiss journeying had provided sufficient material for a 'simple volume' entitled *A Year with the Birds*, which was published in Oxford by Blackwell and appeared as a limited edition, under the authorship of 'An Oxford Tutor.'

With characteristic modesty, Fowler expressed both surprise and delight at the success of this first slender volume. Meeting as it did with acclaim from readers and reviewers alike, there was an immediate demand for a second edition, which appeared, revised and enlarged, later in the same year.

By 1889, as demand for the book continued to rise, its publication was transferred to Macmillan & Co Ltd, and the identity of the author, by then an open secret, was revealed on the cover of the third edition. This was reprinted in 1891, 1902 and 1914 and again in 1925 after the author's death, and finally in 1931.

A Year with the Birds was read and enjoyed by young and old alike. Its aims were modest, its style simple. In the words of its author: 'I have said very little about uncommon birds, and have tried to keep to the habits, songs and haunts of the commoner kinds, which their very abundance endears them to their human friends. I have made no collection, and it will therefore be obvious to ornithologists that I have no scientific knowledge of structure and classification beyond that which I have obtained at second-hand.'

Yet the book filled a very real need at a time when popular ornithology was in its infancy. Indeed, a whole generation of bird lovers grew up under Fowler's influence and several of them generously recorded their debt to him later in their lives. One such admirer was Earl Grey of Fallodon, who in the preface of his celebrated *The Charm of Birds*, stated: 'When I was beginning to notice birds I found delight and help in Warde Fowler's *A Year with the Birds*. Here was a man whose work – he was a Don at Oxford – had, like my own, lain outside the study of natural history. He had been doing for many years with birds just what I was beginning to do: he had found it a pleasant path for recreation. This book did, as it were, blaze a trail, which anyone with an inclination to birds could follow, and thereby be led to find much pleasure.'

A Year with the Birds was to remain the author's own particular favourite throughout his life. He recalled in his *Reminiscences*: 'In this little book there is a certain quality of simplicity and honesty, visible both in the style and the matter, which perhaps is hardly to be found in any other of my books. I had something definite to say in every page, for I had made very careful notes about everything that I did notice, and my memory in matters so interesting to me was very rarely at fault.'

In 1888, Macmillan published the first of two volumes of *Tales of the Birds*, a collection of eight stories, dedicated to Fowler's friend and former pupil, Gilbert Elliott: 'In memory of pleasant days in the sunny summer of 1887.' Intended primarily for children, in the hope that they 'may learn from them to look at a rook or a wagtail with a fresh interest', he said later that although the task of writing the tales had been pleasant, it had been far from easy: 'To be true to the facts of the lives of animals and yet to infuse a few grains of human interest into them, has only been accomplished by one or two writers of real genius... whom I cannot hope to rival.'

Nevertheless, Fowler's stories, which Edward Thomas included in a list of recommended country books in response to an Australian reader, have retained a good deal of their freshness and appeal over the years since they first appeared. The opening story, *A Winter's Tale*, describing the rigours of winter through the experiences of a flock of fieldfares, is both convincing and compelling, providing as it does ample scope for Fowler's sensitivity to weather and to the Cotswold landscape and that of the Marlborough Downs he knew as a schoolboy. Of the other tales, *Jubilee Sparrow* set in Victorian London, displays appealing pathos and *A Debate in an Orchard* is as thought-provoking now as when it was written. Children may still enjoy *A Tragedy in Rook Life* and *The Lighthouse*, but *The Falcon's Nest* appears quaintly irrelevant in a conservation-conscious age and the other stories, though readable, may prove somewhat lacking in vitality for children today. The sensitivity which emerges so clearly in Fowler's bird stories also finds its outlet in his essays. Expressing his loathing for the recently-introduced barbed wire which was beginning to festoon the countryside during the last decade of the 19th century, he speculated that the perpetrators 'perhaps wished to draw the Red-backed Shrike to come and impale his beetles upon those steely thorns.' He went on to lament the suffering that humans and animals too, would be called upon to endure as a result of this remorseless and hideous innovation: '... to put such means of torture upon gates is the cruellest act of all, and not only for human beings ... if horses have a pleasure in life, it is to look over a gate on a

Sunday morning, and rub their necks on the top bar in quiet contemplation of the sheep on the other side.'

Although Fowler had no pretensions as a scientific ornithologist, his ability as a self-taught field naturalist, as revealed in *A Year with the Birds*, and his articles in *The Zoologist*, *The Oxford Magazine* and a growing number of other publications, won him prompt and well-deserved recognition in scientific circles.

Fowler's election as a member of the British Ornithologists' Union led to correspondence with many of the leading figures in contemporary ornithology. During 1889, prevented by the severity of the winter weather from visiting his favourite wildfowl haunts, he contented himself with studying the corpses of the birds on sale in Oxford market. 'Dead ones' he wrote, 'are better than none at all; in one sense indeed a bird in the hand is worth more than many in the bush or on the floods, or even on the shelves of a museum. You can pull him about as you will, you can skin him, and lastly you can eat him.'

Recalling the winter in question, a former colleague wrote: 'W.W.F and I for most of a term dined alone at the High Table in Lincoln. He utilized this to test the edibility of all sorts of birds which he picked up in the market and had cooked for dinner... I remember that not all were equally palatable.'

However, it was the living bird that matter most to Fowler. Writing on 'Birds of the Market' in *The Oxford Magazine,* he lamented at the slaughter of song birds for food: 'Instead of the sparrow, we have the Skylark in countless numbers, yet his flesh is not a whit better than the sparrow's. Observe that he is the skylark in spring and summer, simple lark in winter and at dinner parties, and I really doubt whether we all realise that the bird on the menu is the bird of Shelley and Wordsworth.'

Nor could his examination of the pathetic corpses he purchased be confined to that of the detached onlooker. Describing the colouring on the wing of a drake teal, he wrote: 'I am not thinking of the wondrous green speculum, but rather of feathers which have the broad outer web pure white, each fibre of such transparent pearly delicacy that I can see my hand through the whole, and slashed at the upper end with a broad margin of velvet black; while the inner web is of soft cream-colour, crossed by wavy lines of deep brown.'

Ducks of all kinds delighted him, even those of the commonplace domestic variety: 'Ducks seem to have a more intense satisfaction in the material interest of life than any other birds; and the feature that chiefly expresses their placid enjoyment is that wide laminated sucking-instrument we call the bill.' Later, recalling an incident involving ducks in a ditch at Hinksey, his apt description is laced with typically rich humour: 'There descended with footing slow and sidelong a small company of tame ducks. One frog after another was seized, knocked against the bank to quiet him, held up with the bill erect, his legs waggling in the air, and then quickly swallowed, his progress downwards being clearly visible from outside. No flurry, no excitement, only a quack or two of intense satisfaction, as a thirsty man might draw his breath and say 'Good!' after a long draught of fine ale, and then do it again.'

The beginning of the last decade of the nineteenth century saw Fowler's twin careers as a historian and ornithologist develop significantly. The first of his historical works, *Julius Caesar and the*

Foundation of the Imperial System, was published by Putnam in 1891. In the previous year, he had written what was to become the opening chapter of *Summer Studies of Birds and Books*, a volume which, for its diversity of subject matter, its literary quality and the insight it gives into the workings of its author's mind, merits a chapter of its own.

CHAPTER 8

*We pine for pure air, for the sight of growing grass, for the
footpath across the meadow, for the stile that invites you to rest
before you drop into the deep lane under the hazels.*

SUMMER STUDIES

Summer Studies of Birds and Books was a selection of lectures
and published papers 'written in the leisure of summer days,' revised and sometimes re-written and eventually published by
Macmillan in book form early in 1896.

Dedicated to Oliver Aplin, Arthur Macpherson and Herbert
Playne, three friends with whom Fowler shared many bird-watching expeditions both in Oxfordshire and further afield, the book
serves not only as a record of the author's ornithological observations around Oxford and Kingham, in Dorset, Wales and the
Alps, but also contains a detailed description of his discovery of
the rare marsh warbler's nest in his Oxfordshire village. His
favourite bird family, the wagtails, are given a chapter to themselves; bird song is the subject of another chapter, as are respec-
tively the birds of Aristotle and Gilbert White, whose *Natural
History and Antiquities of Selborne* Fowler was to edit, together
with L.C Miall, for Methuen in 1901.

The one interloper in the book is the author's wire-haired fox
terrier, Billy, whose inclusion Fowler justifies on the grounds that
he 'was all his life in very close relation both with birds and ornithologists.'

Summer Studies was Fowler's penultimate book on birds and the countryside; if we discount two volumes of bird stories, it comprises his one and only major work of the kind published between 1886, when *A Year with the Birds* first appeared and 1913, the year Blackwell brought out *Kingham Old and New*. Unlike the former however, *Summer Studies* made little impact and there was no demand for a second impression. On the face of it, this seems surprising. Both books contain a blend of first hand observation, reasoned scientific theory and literary quality. *Summer Studies* appears to serve as a natural continuation of Fowler's approach to birds and to reveal a logical extension of his thinking and attitude to life generally. Perhaps the basic reason for the difference in popularity between the two books stems from the more scholarly, reflective tone of *Summer Studies*.

A Year with the Birds was essentially an out-of-doors book of general appeal to young and old alike. It appeared at a time when such practical works were few and far between and its fresh, engaging style quickly caught on. *Summer Studies*, on the other hand, requires a more patient determined approach. It is a work of mellow middle-age, a book to turn to and savour by the fireside after a day in the field, rather than a practical stimulus to personal endeavour and as such regrettably failed to capture the public imagination to the same degree as its celebrated precursor.

Yet *Summer Studies*, an intriguing mixture of fact and anecdote, village life and Alpine excursion, scholarship and sentiment, provides an invaluable link between the comparatively youthful phase depicted in *A Year with the Birds* and the mature and accomplished tenor of *Kingham Old and New*, written in the last decade of Fowler's life. The book reveals the author's humanity, perception and intellect – as well as his increasing eccentricity. The similarity with Gilbert White, already clearly evident in *A Year with the*

Birds, takes on a closer, more sympathetic resemblance in the detail of observation and the singe-mindedness with which the task in hand engrosses the mind. But the differences between the two men are also delineated much more sharply. Fowler's intuitive regard for, and interest in his fellow men intensify with the passing years. His circle of acquaintances continually widens; he corresponds with erudite men on a range of intellectual issues; yet the ways of the old village people never fail to interest him and win his affection. Hence the reason why the pronouncements of old Keeper Cook on the role of Wellington in the battle of Waterloo appear alongside quotations form scientific journals and the writings of eminent contemporaries: 'I once asked him if he remembered anything of the Waterloo times. He looked round at me with the one eye he possessed, and said tentatively: 'twas Wellin'ton as won the prize at the battle of Waterloo, wasn't it, sir?' I assured him that his memory had not deceived him. 'Aye', he went on, 'but 'twas old Blucher as done all the vightin'; why, Wellin'ton was a dancing away at a ball till old Blucher come up!' Where Mr Cook got hold of these odds and ends of truth I have no idea. He is now gathered to his fathers, and has vanished away from us like the smoke.'

Gilbert White, of course, confined his observations almost exclusively to the sphere of natural history. Fowler, on the other hand, is drawn to the human condition, in whatever state he finds it and sees no need to separate his observations on his fellow men from those on lesser forms of life. This approach emerges fully in *Kingham Old and New*; in *Summer Studies*, however, the trait is clearly established.

Always thorough in his ornithological studies, as in the other branches of scholarship in which he involved himself, Fowler's humility and readiness to learn showed no sign of diminishing with increasing age. He made little effort to conceal his impatience with the hastily-drawn conclusions of less-painstaking observers, however, such as the 'certain popular writer who (though it is but lately that he learnt the difference between the Grey and the Yellow Wagtails) assures us boldly that the object of the motion (of their tails) is to aid the bird in balancing itself.'

Fowler also questioned, in milder tones, W.H. Hudson's lamentations on the depletion of the larger birds of prey in Wales, instancing his own contrary observations, which suggested that despite the depredations of egg-collectors, the larger predators could still be found in reasonable numbers if persistently searched for. Acknowledging the value of Hudson's plea for protection for these birds, Fowler confessed to finding Hudson's words 'irritating... to those who have lived all their lives in England' – clearly a reference to Hudson's South American origins and perhaps a dubious, if not unfair argument!

The two men also differed on the question of personal preference concerning birds. In *Summer Studies*, Fowler stated the case for 'birds that can be perused; not hasty ones that are up and away the moment they catch sight of you, nor huge ones, such as Mr Hudson loves, sailing solemnly over your head and vanishing over the hill while you adjust your glass.'

In his *Birds and Man*, Hudson took up the challenge: 'It is only natural, in an England from which most of the larger birds have been banished, that he (Fowler) should have become absorbed in observing and in admiration of the small species that remain.' Hudson went on to confess a distinct preference for larger birds, a clear association with his own experience on the pampas of Argentina.

The two writers were at one on the appreciation of bird song, however. Hudson quoted Fowler liberally and with obvious agreement on the question of the distinctive quality of the songs of the chiffchaff and willow warbler, and on the interpretation of warbling generally (in *Adventures among Birds*).

It was Fowler's work on the marsh warbler, however, that Hudson, like so many contemporaries, was quick and generous to acknowledge. Fowler began his observations on this hitherto little known summer migrant in Oxford during the summer of 1888 and continued studying its behaviour, first in Switzerland and later in his adopted village of Kingham, until 1906. He was the first naturalist to attempt to describe its remarkable song – 'a very sweet silvery individuality... which makes it quite unmistakable' – and his notes on its powers of mimicry and on its nesting habits were published as a paper by Blackwell and read by Fowler to the Oxfordshire Natural History Society in November 1893.

Of the three ornithologist friends to whom *Summer Studies* is dedicated, and whose fellowship Fowler shared on bird-watching expeditions both in England and Switzerland, perhaps the one with whom he had the greatest affinity was O.V. Aplin (1858-1940). Himself an Oxford man, Aplin lived for most of his life at Bloxham, near Banbury and was the foremost authority on the birds of Oxfordshire. His book on the subject, published in 1889, was a model for other aspiring writers being based on first-hand observation and revealing a sound knowledge of the geographical and geological factors governing the distribution of species. Aplin openly acknowledged his indebtedness to Fowler, especially concerning the birds of west Oxfordshire. He was a frequent visitor to Kingham and the friends walked for miles around the area on bird-watching expeditions. Fowler acted as best man at Aplin's wedding and was godfather to Aplin's son Robert, who recalled being given a golden half-sovereign on one of Fowler's visits to Bloxham.

Fowler had the greatest respect for Aplin's abilities as a field naturalist. In 1891, he persuaded his friend to accompany him to the Alps, the idea being that the two men should combine their skills in the search for birds: 'I was to find out the localities in which we were likely to come upon new or interesting birds, and my friend was to bring his accurate scholarship to bear upon them when found.' The partnership met with immediate success. At the end of the long street at Interlaken, leading towards the lake of Brienz, Fowler proposed that they should follow a footpath between the river bank and a tract of damp scrub. 'My friend plunged into it, while I went on a little further. Almost directly he called me back, and by the waving of his umbrella I saw that he had made some discovery. It was indeed a discovery; it was the next of a Marsh Warbler.'

In the following year, 1892, Fowler was to summon Aplin to Kingham to confirm the presence of the marsh warbler singing in an osier bed within ten minutes' walk of his home. The bird was to be the object of Fowler's undiminished interest over the next fourteen years, culminating in the eventual desertion of the nest site, possibly arising from the clearing of the osier bed.

Arthur Holte Macpherson, the second of Fowler's friends to whom *Summer Studies* was dedicated, was close to Fowler for much of his life. Like Aplin, he was a good field naturalist and it was after consulting him that Fowler, with considerable misgiving, decided in June 1893 to take the nest and eggs of the pair of marsh warblers which had built in a vulnerable position in a small osier bed 'partly because it would be no longer safe from the village boys, as I was about to go abroad.'

Fowler's action appears inconsistent to the modern conservation-conscious mind, especially when his reputation rests on the study of the living bird, rather than the skins on which most of his contemporaries were content to concentrate. However, it should be remembered that at a time when modern scientific techniques and methods of recording were unthought-of and bird-nesting was a natural pastime among country boys, Fowler's dilemma was a very real one: 'I was sharply criticised for taking it,' he wrote several years later, 'but I was confident that the birds would build again, and it turned out that I was right. They returned to this spot the next year.'

The third friend, Herbert Playne, a teacher of mathematics at Clifton College, and a contributor, like Fowler, to *The Zoologist*, became an authority on the birds of the Bristol area and accompanied Fowler, Aplin and Macpherson on several bird-watching tours to Wales and to the Alps. Playne related how on one of their long Swiss train journeys, their peace was disturbed by 'a very voluble French lady, who chattered incessantly far into the night'. Fowler, Playne went on, 'leant across to me and said 'Sedge Warbler'' – a bird noted for its noisy nocturnal habits.

Although Fowler devoted countless hours to the study of the marsh warbler and other small song-birds, the pride of place in his affections undoubtedly went to the wagtails: 'It is impossible ever to weary of wagtails,' he wrote in *Summer Studies*, 'we are never altogether without them, yet whenever they present themselves to us we are constrained to give them our attention... There is a never failing pleasure in contemplating their symmetry of form, their beauty of colouring, their graceful flight, their unobtrusive confidence, and that constant unresting activity of theirs – an activity which some mysterious grace of mental build never suffers to degenerate into fidgetiness.'

It is hardly surprising that such exquisite little birds appealed so much to a man of Fowler's sensitivities. His chapter on the wagtail family in *Summer Studies* is filled with observations, theories and happy reminiscences, recalled, many of them, from distant memory and recorded with obvious delight. The yellow wagtail, in particular, enchanted him: 'So light and sylph-like are they that the (wheat) stalks were hardly bent beneath their weight; and I could not help singling out one of these on which a bird had been resting, and trying to measure with the touch of my finger the weight of that fairy figure.'

Fowler spent part of most summers during his middle years in Dorset and the natural glories of that county held him in thrall. He had met the Dorset poet William Barnes during his time at Marlborough and held his verse in high regard, so much so, that he set the lines *In the Spring* to music and often played the piece to his friends. He also compiled a paper, *Dorset and its Poet and Novelist*, a further tribute to the dialect poet.

One particular entry in Fowler's diaries relates to the train journey to Dorset and includes colourful descriptions of his fellow travellers:

'1. Kingham to Oxford – a very bulky red-faced red-bearded good-humoured man sat in one corner: what struck me was that he took special pleasure in contemplating his gloves, with which he vainly endeavoured to cover two enormous red paws.

At Charlbury there got in a pleasant-faced gentlewoman of the country, with a Wessex accent; her husband was a Blue-ribbonist and the driest of men. I am sure he was a Scotchman; ... He was unbearable: cold, virtuous, feeble, dull – and I don't see the good of commemorating him further.

2. Oxford to Didcot – King Henry VIII sat opposite me; see his picture in Ch. Ch. Hall. I identified him by his short bluff sandy all-round beard.

3. Didcot Station – Superintendant Head of Oxford, and three small boys in his charge: possibly going to a reformatory, but they were treated largely to pears and buns by the people in the waiting-room. In this melancholy place a young woman sat near me, who was always suddenly getting up and knocking or kicking the seat-legs. As I had a headache, I hated this jerky young woman.

4. Didcot to Swindon – Nobodies – A very mild gentleman passed a few remarks with me, but seemed in a very hopeless position as to his whereabouts and intentions.

5. Swindon to Trowbridge – Aggravated by a man of the famer type, who began by eating a cold chop which increased his previous greasiness; he ate it off the bone, and then holding the bone with one hand at each end, he licked away at the middle. Then he got greatly excited about his old home, near which we were passing: and pointed out all the localities. Finally he shut up both windows, because his wife had the toothache: which caused me to leave the carriage at Trowbridge.

6. Trowbridge to Dorchester – Working men or railway servants. I noticed that when asked a question they always at first affected not to hear, and called out A-a-y in a loud tone, this I believe is a common device to gain time in rustic conversation.'

The Dorset holidays usually consisted of a fortnight's stay at West Lulworth, accompanied by an old pupil, R.S. Osler – 'a very pleasant companion, whose natural reticence quite suited my mood.' They spent a great deal of time wandering over Bindon

Hill, hallowed ground almost to Fowler, and the subject of one of the chapters of *Summer Studies*: 'I often doubt whether there can be such another hill as Bindon in the islands; I at least have never found it.' The literary quality of *Bindon Hill* was of the highest order; Fowler is said to have told a friend that it was the only piece of writing on which he had taken pains with the style. As an all-round descriptive essay on a feature of the British landscape it is certainly memorable: 'To understand him truly it is not enough to contemplate him from without. You must spend whole mornings with him, lying on him, and being of him. Better to be bookless there, in my opinion, even on the warmest day; I cannot keep my attention on the page, there is so much life and fragrance around me... whether I turn southward to the sparkling sea with its white sails, or look northward over long miles of purple plain, or lie down and look into the long dry grass which the sun is turning golden, and catch the millions of gossamer webs, stretched by some invisible fairy spider from blade to blade over the sward.'

The field of classical scholarship was to claim Fowler's literary energies after the completion of *Summer Studies* until the writing of *Kingham Old and New* towards the end of his life and it is to that demanding yet productive phase that we now turn.

Corner House, Fowler's first Kingham home.

Fowler House. Warde Fowler's home until his death in 1921

Kingham station. Chipping Norton Junction until 1909

Another view of the station used regularly by Fowler.

The new lychgate, 1912. Fowler centre, holding book.

Warde Fowler's grave

The former rectory, Kingham

Kingham green, early 20th century

The Castle or Monte Rosa, John Barrow's home.

The former village school, demolished in 1911

The River Evenlode at Kingham

The osier bed in which Fowler found the
marsh warblers'nest

The road off Church Street named after William Warde Fowler.

Bryn Cwrt, Llandovery, Fowler's Welsh holiday destination.

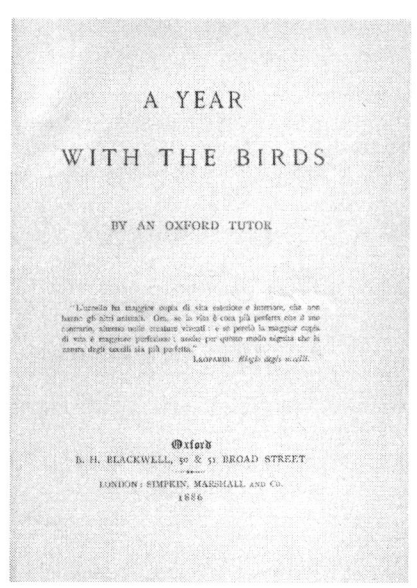

'A Year with the Birds,' Fowler's first book,
published anonymously.

Lincoln College tutors, 1887. Warde Fowler in the middle.

CHAPTER 9

History is like Natural Science in this, that when once man's curiosity is excited in any department of it, he will always find work to do and work that may be really worth doing. Wherever he moves, he will find himself saved from that intolerable ennui that men feel who have never learnt to keep their eyes open.

SCHOLAR AND VILLAGER

The last years of the 19th century saw a strengthening of the bond between Fowler and the village of Kingham and the beginnings of a corresponding loosening of his attachment to Lincoln College, which was to culminate in his resigning from the Sub-Rectorship in 1904.

Fowler often said that he could never achieve any worthwhile writing in Oxford. Kingham on the other hand, afforded him the peace and seclusion he needed to compile the works of scholarship upon which he wished to devote the remainder of his life. His deafness and other personal circumstances played their part in resolving this matter. His father's death in December 1899 resulted in the final break-up of the family home in Wales. Fowler's half-sisters, Florence and Emily, went to live at Malvern and his sister and inseparable childhood companion, Alice, came to share his home at Kingham.

Although brother and sister had kept in constant touch throughout William's college and university days, Alice's coming necessitated certain changes in her brother's lifestyle, as he recalled somewhat wryly in his *Reminiscences*: 'As I had been a bachelor in all my ways it was necessary that the house should become, as we called it, 'ladyfied', and the process, if not particularly agreeable, was anyhow successful. Our drawing room and dining room became presentable, and the study was kept in its old condition as a working and smoking room. My Oxford guests continued to come as before, worked and smoked, and my sister, in spite of her deafness, entertained them with her own peculiar sweetness and intelligence. Owing to her cheering companionship, I was able to do more good work in the way of writing books than I had ever done before.'

Nevertheless, Oxford still retained a firm place in his heart and he spared no effort in striving for the preservation of its cherished heritage. In 1901, after hearing of a proposal to extend the University science buildings into the Park, he wrote as Sub-Rector of Lincoln in a privately printed open letter to the Vice-Chancellor, his old friend Thomas Fowler: 'I feel it is incumbent on me to make my views known to you and to others, since in this matter I have not only a personal responsibility, but (if I may say so) a sufficient knowledge of both sides of the question... No-one knows better than I that Science will be starved at Oxford unless it can find room in the coming century for many new buildings... but at the same time I am sure that no-one understands better than I do the inestimable value of an unmutilated Park.'

Thirty years' experience of the examination-orientated structure of the British university system of his day prompted Fowler to question seriously the validity of what constituted a university education. In *An Oxford Correspondence of 1903*, published by Blackwell, he voiced his views on the subject through Edward Slade, a fictitious don, in an exchange of letters with his pupil, Jim Holmes, based on Fowler's own pupil and friend Horace Mann: 'We in England have become so completely salted, soused and pickled in these exams that we no longer use our natural intelligence in judging of them – we take them for granted, and never or rarely enquire into their effect on the human mind.' Throughout his life, Fowler drew no distinction between work and leisure. The pursuit of scholarship for its own sake was to him the obvious and most important ingredient of a university education and he saw an over-emphasis on examinations as a threat to this ideal: 'Our system of examinations has seriously damaged the natural intelligence of the nation by almost destroying the freshness of interest which a fair average of boys ought to take in their work, and by robbing them of much mental freedom and elasticity.'

He remained unconvinced that sporting activities were the natural means of spending leisure and pleaded the case for boys to be free to pursue their own interests, as he himself had done, rather than having their time over-organised: '...The weariness of work... has... exaggerated the importance of athletics as a relief from tiresome routine.' His views on the role of tutor were equally challenging: 'A tutor's duties are bracing, humane, social, it is only now and then, when a talkative bore descends upon you, that you are tempted to wish that he might collapse upon your hearthrug, smitten with some fell disease.'

The relinquishing of the Sub-Rectorship did not mean the end of Fowler's attachment to Lincoln College. He continued to receive pupils at Kingham until 1908 and resumed this activity again during the war years, until 1917, when the death of his sister finally prompted him to sever this link. He remained a Fellow of Lincoln College until his death, however, maintaining all his rooms at the college and retaining an interest in college affairs until the end.

The year 1904 saw Fowler, freed at last from 'the everlasting necessity of making an income by routine college work,' embark on the belated career of concentrated scholarship to which he had so long aspired. Scarcely a year passed without some new contribution from his pen to the field of ancient history. His books and papers were acclaimed by scholars and critics alike. B.W. Henderson, the Roman historian, called him 'the greatest ancient history tutor of his generation.' Professor R.S. Conway wrote: 'Warde Fowler's work was characteristic and unique.' Professor H.J. Rose called him 'the best type of Victorian don.' Of his status at Oxford, Sir Samuel Dill wrote: 'He is the greatest name among Lincoln men of his time.' 'What made his works so readable as well as informative,' wrote Dr. V.H.H. Green 'was his power of sympathetic interpretation.' However, in expressing his own contemporary view that Fowler as a scholar lacked skill as a textual critic, Dr Green drew attention to Fowler's single brief visit to Rome as a young graduate, referred to earlier. Fowler had acknowledged this seemingly inconsistent attitude to history in *A Correspondence of 1903*: 'I have a foolish but constitutional dislike of ruins, broken and disjointed things that make not the great qualities of mankind, but the weak ones.' This ambivalent attitude to the physical origins of his classical scholarship persisted, even when his studies had virtually become his life's work.

In 1905, he was invited to accompany some friends on a short voyage to the Greek islands in the steam yacht *Argonaut*. While the others devoted themselves to the glories of the Grecian past, Fowler went off in search of birds 'of which I everywhere found one or two new ones... at Delphi the Orphean warbler, the Ortolan bunting and the Egyptian vulture... and so to Ephesus by rail. Here for natural history alone we might have spent a week, and I was of course obliged to pay due respect to the antiquities.' Fowler leaves us in no doubt where his interests lay as he went on: 'So little time was allowed us for these (the antiquities) that I was content to listen to a lecture, and as we had on board both Bryce, Macan and other scholars and historians, I was able to rest quietly while they discoursed.' It is entertaining to speculate what the others thought of the deaf, grey-haired classical scholar whose idea of savouring the delights of ancient Greece differed so radically from their own. A full account of Fowler's ornithological observations on the cruise was published in *Macmillan's Magazine* later in 1905, under the title of *Bird Life in Greek Waters*.

Fowler's friend, Horace Mann, left an intriguing description of Fowler's appearance and behaviour during a holiday they spent together at about this time: 'He was not an ordinary figure at any time; on a journey or holiday he must have been very difficult to classify at first sight. He looked old and bent, and yet he could be, and usually was, very active and quick in moving... He was obviously an English gentleman but his hair was rather long and his soft felt hat was worn with a full dome. He appeared to be very deaf, and yet he clearly understood. He looked absent-minded, and yet he had noticed. I suspect that he had an almost impish joy in bewildering serious people. A fellow traveller, finding that he came from Oxford and was a professor, asked him a string of what he thought tiresome questions: 'Had he known Matthew Arnold?' He said 'Mat Arnold? Oh yes, I sat next to him once at dinner. But he would not have anything to do with *me*.' Then, laughing: 'He parted his hair in the middle.'

Fowler's habit of wearing a soft felt hat with the dome at the full gave rise to the story that he once arrived at Kingham station wearing three hats superimposed on one another for convenience of transport.

His longstanding disappointment at the limited recognition his scholarship had received over the years from his own university was to some extent compensated for when in 1909 he received an invitation from the University of Edinburgh to give a course of lectures on Roman religion for the Gifford Foundation. These were such a success that he was asked to deliver another series in the following year. In 1911, the lectures were published in book form under the title *The Religious Experience of the Roman People*. Other honours came Fowler's way at this time and all were accepted with characteristic modesty. His portrait, which still hangs above the High Table in the Common Room at Lincoln College, was painted by Macdonald of Winchester. Two honorary degrees, a LLD at Edinburgh and a LittD at Manchester, were conferred, the citation to the latter referring to 'a scholar in whom the students of his writings gratefully recognise perhaps the truest living embodiment of the ideal of humane letters.' Recalling the occasion in a letter to a friend, Fowler wrote: 'My hood is so magnificent (it is the only finery that I bought) that I can imagine it decorating some Cretan swell in the prehistoric age. My sister thought it would do nicely for our niece's new baby.'

By now in his mid-sixties and, like his sister, no longer capable of strenuous walks, Fowler used his lecture fees to buy a comfortable four-wheel carriage, referred to as *The Gifford*, which was garaged by the village carrier and cab proprietor, Robert Brick, who provided a horse and acted as coachman when Fowler and his sister, accompanied often by friends or relatives, set off on an afternoon drive to the Barringtons or Bourton-on-the-Water.

The year 1911 saw the passing of two of Fowler's best-loved friends, Sidney Irwin, one of the *Two Ideal Schoolmasters* of his *Essays in brief for Wartime*, and Samuel Davis Lockwood, Rector of Kingham, to whose memory Fowler was to dedicate the book on his adopted village upon which he was now working. Increasing deafness was by now affecting brother and sister alike; special attachments were fitted to their pews in church to accommodate their ear trumpets. Ernest Lainchbury, in his book *Kingham, the Beloved Place*, states that Fowler, who gave up the churchwardenship in 1913 after fifteen years' service, showed his disapproval of the lengthy sermons of the new rector by reading a book during their half-hour duration. This hearing disability did not prevent the Fowlers from enjoying their music making, however. The novelist Anne Douglas Sedgwick, who had married Basil de Selincourt and come to live at Kingham in 1909, gives an intriguing description of a Christmas party held at her home in 1912: 'Yesterday we had our Christmas tree – Lucinda Smith and Dorothy Hills and the Belcher children and my washerwoman's children, and two dear little things who came in the summer to pick berries for jam for us – 9 children in all; and dear old deaf Warde Fowler, our Oxford don here, and his sister, who, after the tree and the presents, played us duets quite delightfully – Brahms: Scherzo from the 3rd Symphony; I think they enjoyed it all as much as any child here.'

Fowler's involvement in village life was both far-reaching and profound. In addition to serving as churchwarden, he was a manager of the village school for over a quarter of a century and a tireless campaigner for the well-being of the village children. His former pupil and close friend, Horace Mann, had become a member of the newly-established Board of Education in 1904, and Fowler wrote to him at length, pointing out some of the problems the new Education Act was causing in the management of village schools: '... When men get together in Government office there is no saying what deterioration they may let suffer. Certainly the office has been bullying us almost to madness... The inspectors... turn up every other week and threaten us. We are writhing under this tyranny which has increased tenfold since the new Act came in.' As correspondent manager to the school, Fowler was a frequent visitor and knew all the children and their parents. In June 1910, while he was giving Gifford lectures in Edinburgh, a violent thunderstorm broke over Kingham and Fowler, always intensely interested in the phenomena of weather, was delighted to receive details of the storm in letters from the schoolchildren. Later, he used these letters as evidence of the children's ability in writing and spelling, to confound a cynical school manager who had asserted that the children were badly taught: 'I have heard no complaints of this kind since then.'

Great was Fowler's delight when in 1912 the decrepit old church school, which had formerly served as the Rector's tithe barn, was replaced by a spacious new council school. Fowler was the unswerving ally of the teachers in their efforts to widen the curriculum and to incorporate new ideas into their teaching. He was especially concerned that nature study should be taught by teachers who were themselves well-versed in the practical aspects of the subject: 'The only person who can really help the children in these things is one who is learning himself all the time and learning not only from books but using them just as a help.' In *A Few Last Words*, the concluding chapter of *Kingham Old and New*, written just after the new school opened, he admonished those who considered that the task of the school lay merely in providing a traditional basic grounding: 'They are utterly wrong in supposing that in these days the three R's are all that is necessary; that might have done well enough in the days of the yardlanders, but not now... School life can be made a real delight to the children... The brainless, or the brain-destroying, order is passing away.' Looking forward to thirty years hence, he predicted: 'Our schools will be more really full of life and freedom, and the parents will have begun to believe in them.'

The man whose vision and intellect had already indicated the way forward in the higher realms of learning offered a message equally relevant and essential to those concerned with the education of the youngest children in a humble village school.

CHAPTER 10

Most of our villages have some substantial tale to tell, some small mite to cast into the treasury of our ancient history. I believe Kingham to be one of those.

THE GILBERT WHITE OF KINGHAM

It has already been noted how William Warde Fowler followed in the tradition of Gilbert White in recording with care and devotion the natural history of his village. Although essentially a field naturalist, relying for the greater part on his own painstaking observations, Fowler had the benefits of a hundred years scientific advancement denied to the celebrated pioneer in the field of local natural history. Yet despite the fact that Fowler's writings, concerned as they were with the place of man in his rural environment, inevitably cover an aspect of village life outside the scope of White's *Natural History*, the parallel is inescapable, and *Kingham Old and New* deserves a place alongside *Selborne* on any country-lover's bookshelf.

This affinity between the two naturalists had been remarked upon as early as 1893 by E.D. Lockwood, brother of Rector Lockwood of Kingham, in his book *The Early Days of Marlborough College*. Himself a naturalist of some repute, Lockwood had settled in Kingham after retiring from the Indian Civil Service. Writing of his village contemporaries, he remarked: 'We have our Gilbert White in my neighbour, Mr Fowler, the historian and fellow of Lincoln College, Oxford.'

Writing over half a century later, David Green (*Country Neighbours*, Blandford, 1948) considered Fowler's books to be 'not nearly so well known as they deserve, though those of us who have them are inclined to speak of them so well that our copies are constantly on loan to the uninitiated, so much so indeed, that I sometimes think that it cannot be long before the delightful old don is rediscovered and made much more of and accorded the niche that must surely await him not so very far from Gilbert White's.'

Fowler's admiration of White knew no bounds. In his chapter entitled *Gilbert White of Selborne* in *Summer Studies of Birds and Books*, he describes the Hampshire clergyman as 'the first country gentleman who could shake himself free from the tyranny of books, and describe what he saw around him in simple and engaging English.' He goes on to quote with obvious approval White's stated belief that: 'Men that undertake only one district are much more likely to advance natural knowledge that those that grasp at more than they can possibly be acquainted with; every kingdom, every province, should have its own monographer.'

Temperamentally it seems, the two men were very much akin; certainly the passing of years bore witness to Fowler's increasing desire to live the kind of life to which he attributed White's success as the incomparable monographer of English rural natural history: 'For such close and keen observation as his, it is really necessary to be master of one's own time, to be absolutely free from hurry and interruption, and this not only that a bird or insect may be carefully watched, but that what is seen may sink quietly and surely into the mind... The unique value of his book is mainly due to the persistence with which he followed his own instinct, and to the complete ease and isolation in which his acute mind worked.'

Unlike Gilbert White's observations on his beloved village, Warde Fowler's writings on Kingham have never been collected into a single volume. This seems a great pity, not merely to afford an easier comparison with *The Natural History and Antiquities of Selborne*, but because such a collection would provide a unique survey of the history, natural history, weather and personalities of an Oxfordshire village over a period of half a century.

Obviously, *Kingham Old and New*, an excellent village monograph in its own right, would provide the basis for such a compilation; reference to several other sources would, however, further diversify and enrich the work and produce an even more comprehensive and rounded picture.

Those additional sources are many and varied. They include *A Year with the Birds*, whose chapters *A Midland Village – Garden and Meadow* and *Railway and Woodland* deal with the bird life and habitat of the Kingham area, and *Summer Studies of Birds and Books*, in which four chapters are devoted largely to the same subject. In *Kingham Old and New*, Fowler himself draws on his observations in *A Year with the Birds* to show changes in the distribution of certain birds, such as the redstart, yellow wagtail and willow warbler, that had taken place in the quarter of a century separating the two works.

As well as his bird books, Fowler's articles in various learned journals provide a considerable amount of additional material relating to Kingham, especially the long essay *A Ridge of the Cotswolds*, originally written for the *Oxford Magazine* and later published, along with two other essays from the same journal, in R.T. Gunther's *The Oxford Country*, which was issued by John Murray in 1912.

It could perhaps be argued that some of the *Tales of the Birds*, set faithfully and evocatively in and around Kingham, would be worthy of inclusion in a collection of writings on the village. *Selena's Starling*, for example, though ostensibly the story of a fairly unremarkable bird, is in fact based on an incident in the life of Lucy Porter, a village character of the kind whom Fowler found intriguing.

But it is in the richly diverse and extensive amount of unpublished writing – letters, diaries and entries in Colonel Barrow's vast 'log' – that we find the most valuable and rewarding accumulation of supplementary material on Kingham and its natural history. For Fowler was prolific, both as a letter-writer and diarist. His frequent visits to the Colonel's residence, too, were invariably commemorated by an entry in the celebrated 'log'. Although the greater part of Fowler's diaries have long since been lost, Professor R.H. Coon went to great pains to contact those of Fowler's correspondents still living at the time of his researches for the biography in 1933 and their recollections of Fowler, together with his letters, add considerably to our knowledge of village life at the time.

It should also be borne in mind that a good deal of Fowler's writing on natural history had an authority and significance well beyond his adopted village in the best tradition of Gilbert White. His observations on the marsh warbler, for instance, were the most detailed of his time, appearing in *The Zoologist* before being printed in booklet form by Blackwell. His work on bird migration, published also in *The Zoologist*, added to the sum of knowledge on the subject being accumulated at the time.

Fowler's interest in the phenomena of weather and his literary skill in describing it led to the inclusion of his essay *In Praise of Rain* in *The Book of the Open Air*, compiled by Edward Thomas and published in 1907. Thomas is said to have told Fowler's friend Horace Mann that he knew of no-one with a finer sense of the characteristics of the British countryside. Although Thomas studied at Lincoln College during Fowler's time, there is no actual record of their friendship. It is worth noting that one of Thomas's most well-loved poems, *Adlestrop*, owed its creation to the poet's train journey between Oxford and Ledbury, and in particular to the halt at Adlestrop Station, which until 1964 was the next stop on the up line beyond Kingham.

However, in a letter to his friend Eleanor Farjeon, written in April 1914, Thomas, in accepting an invitation to join a small group of friends for a short holiday at Kingham, wrote, 'You will like Kingham which I know a lot about, because a former Sub-Rector of my college lives there (and wrote a history of it later) Warde Fowler.' Like Thomas, Fowler was an observant train passenger. According to Mann, 'Whenever there was anything interesting to see, he managed to be looking out of the carriage window.'

An entry in Fowler's diary for 1884 bears out this observation: 'I am always ready to use the train, to save my legs and reserve them for better exercise than pounding along dusty roads, and I am strongly inclined to think that a railway cutting or embankment gives life and variety to the scene, especially when you now and then see a column of clear pure snowy steam stealing on past wood and meadow and feel that it connects you with the life of the world.'

Fowler's letters reveal his breadth of interest and his ever-enquiring mind. He often lamented at his lack of scientific education but never allowed these shortcomings to inhibit his determination to solve the problems he set himself. There is rather more than a superficial resemblance to Gilbert White in his correspondence with his friend and former pupil, the scientist N.V. Sidgwick in 1905: 'In that same fog before Christmas, the air being absolutely still, I was coming up from the station here, and on approaching the bridge which carries the road over the railway, I saw the telegraph wires... swinging up and down in so curious a way that I thought for a moment I must have got a bilious attack or something wrong with my eyes... Rime was on the wires, but would that account for it?'

Three years later, writing to the same correspondent, it was the question of a pig's reaction to sulphate that concerned him: 'Can you tell me whether it is a fact that there is a deposit of sulphate on the rails caused by the friction between the solid rails and wheels going over them? ... This morning I assisted (by request, as having an umbrella) in driving an old sow over a level crossing... the keeper... said that 6 out of 10 pigs display the same feeling; in other words, something in the metals is taboo to them; I saw her smelling them with obvious disgust. The man explained it intelligently as sulphur deposit... I am going to enquire further into the experience of pig drivers.'

But it is to *Kingham Old and New* that we must turn to find Fowler's greatest contribution to the literature of English rural life and that aspect of his output closest in spirit to White's *Selborne*. Published by Blackwell in 1913, in his sixty-sixth year, it was begun three years earlier, the germ of the idea originating from an article on the economic history of Kingham for an American magazine which Fowler had written several years before. 'It is a nice-looking, strongly bound book,' Fowler wrote in his

Reminiscences, 'and will last exactly as it is for generations.' Nevertheless, he could not hide his disappointment that the demand for the book did not necessitate the printing of a second edition, nor that the village people were reluctant to part with five shillings to buy a copy: 'They all want to read it, but as long as they can get a chance of borrowing it they will not come out with their money.' Reason rather than rancour prevailed however, prompting him to add: 'It is a pleasure to the author to give away books that he knows will be treasured up and they may be right in presuming that they have a certain claim on the author.'

The reviewers, however, were quick to recognise the excellence of the work: *The Times* acknowledged the parallel with White's masterpiece: 'The Kingham observations are in the right succession of Selborne.' *The Athenaeum*, too, found: 'In the beautiful limpidity of his style and the mingling in him of the antiquary and the naturalist he often makes us think of Gilbert White.' E.V. Lucas in the *Spectator* wrote: 'This book is human to the core. And more than merely human, it has personality and an underlying tenderness and sense of the best in life that makes it literature... Between the lines on every page one catches glimpses of the one who loves his fellow men, and has acquired rich stores of sunny wisdom and sympathy from an observant life of tranquil delight in nature, books and neighbours.'

What of the book's status at the present day? As Fowler predicted, its excellent binding has ensured its survival. He would have been immensely gratified to know how the book has been treasured by successive generations in Kingham itself; the present writer was shown an obviously cherished copy by a farmer soon after moving to the village and later learned of the existence of several other copies, handed down through generations of village families.

Despite certain inaccuracies since revealed by later studies of Kingham's early history, the book has lost none of its appeal today. The three opening chapters dealing with the history of the village from its earliest origins to the enclosure of its open fields (which did not take place until 1850) remain still essential reading for anyone wishing to understand the factors governing the patterns of settlement and early agrarian techniques associated with Kingham and its environs.

The chapter on *Old Village Folks* shows Fowler's humanity - and humour - at their best, a delightful glimpse at the lives of a handful of Victorian country characters, who but for Fowler's skilful and sympathetic pen, would have disappeared without trace into oblivion.

Birds, understandably, have a chapter to themselves and also command a considerable share of the chapter entitled *Curiosities of Coxmoor*, a village meadow with an even richer store of natural history associations for Fowler than the Yantle, which featured so prominently in *A Year with the Birds*.

There is also a most valuable chapter on flowering plants, a subject on which Fowler, despite his modesty, was highly competent, and whose own knowledge was supplemented by that of his sister, Alice. No flower, however humble, is considered unworthy of inclusion in this chapter and the butterflies associated with the flowers in their favoured habitats are also described.

Fowler indulges to the full his fascination with the vagaries of the weather, devoting no less than three chapters to the subject – one to the incredibly violent thunderstorm of 17 June 1910, which to his regret he missed, being in Edinburgh delivering the Gifford lectures; one to the great drought of 1911, and a further chapter to the severe snowstorms of 1881 and 1908. Any fears the reader may have that these chapters lack appeal are instantly dispelled as Fowler chronicles in his characteristic engaging style the effects the extreme weather conditions had in the countryside generally and on its human and animal inhabitants in particular. Describing the thunderstorm, he records with obvious relish a letter he received in Edinburgh from one of the village schoolchildren: '... a quiet boy who is rather a favourite of mine; he is not so brilliant a writer... as will be seen by the following fragment but it contains a valuable fact.' He goes on to quote: 'The lightning struck the Manner House and knocked a lot of slates off of it. When it was thundering and lightning the cuckoo was singing all the time.'

Kingham Old and New concludes, as we saw earlier, with a chapter entitled *A Few Last Words*, in which Fowler argued cogently and with absolute conviction for an improved way of life for the village people, especially the children, with whom he lived. He declared himself firmly opposed to the ideas current at the time for the throwing-off of the feudalism of the landlords in favour of a system of peasant properties: ' If our economic system is to be changed, the new one must be nursed in some way that will save us from intellectual destitution.' He saw education as having a major role to play in the future well-being of the village community and worked to his utmost to bring about that happy state of affairs.

An event of profound and far-reaching significance was about to break on the world however, sweeping away in its relentless tide many of the men, and values, that Fowler held dear – the First World War.

CHAPTER 11

It is quite in keeping with the tendencies of the time that I am unlearning in my old age some of the settled conclusions of my younger days.

THE LAST YEARS

'It may be difficult at a future date,' Warde Fowler wrote in his *Reminiscences*, 'for some people to understand how it could have been possible even for persons far beyond the fighting age to be so distracted by the daily war-news as to make it impossible... to bring the reasoning powers to bear upon other matters.'

To a man of Warde Fowler's sensitivities, for whom the Boer War had brought deep distress culminating in insomnia, the Great War came as a numbing, almost heartbreaking catastrophe. For a year and a half, he was unable to concentrate on any serious work and spent his time seeking what solace he could from re-reading the plays of Shakespeare, the novels of Sir Walter Scott, Jane Austen, George Eliot and Dickens as well as the works of Darwin and Wallace. Music-making with his sister also gave him some comfort, especially that of Mozart, on whose works he had written a ninety-page privately-printed paper, *Stray Notes on Mozart*, four years earlier.

By 1915, the heavy price of war had begun to exert its grim toll on Kingham. News of the killed and wounded reached the village and Fowler, for whom letter-writing had once been a natural and pleasant activity, was prompted to embark with a heavy heart on 'one of the few duties with which I could charge myself – to write letters of sympathy and condolence for those who were in trouble in the village.'

Sadly, his dismay at the war, intensified by the severing of his contact with German scholars and disillusionment with those of their number who appeared to him to condone their country's belligerent conduct, led to an uncharacteristic bitterness which for a time soured his relations with both friends and villagers alike. Because of his strong objections concerning the proposal to remove an ancient decorated tomb from the chancel of the parish church, the patron, C.E. Baring Young, withdrew his offer to complete restoration work to the building. Fowler also took issue with George Phillips, a yeoman farmer, fellow school manager and good friend of many years' standing, over the latter's decision to allow a group of corrugated bungalows to be erected on his land on the edge of the village, vowing to avert his gaze from the sight whenever he passed! Yet even at this stressful phase of his life, Fowler's letters reveal traces of the distinctive sense of humour which endeared him to his friends. Writing to the editor of the *Classical Review* in 1915 concerning the forthcoming publication of his notes on Ovid, he anticipated criticism from an eminent man of letters: 'Do you think Housman the Terrible will eat me? I don't know him personally but I am told that those who cross his path are never seen again, or only a trifle of bloodstained relics of them.'

But his innate humanity and tolerance of spirit eventually triumphed over this adversity. In 1916, he wrote to a friend: 'I was afraid my indignant letter might have disturbed you. It was one of several I wrote to my friends, wishing them to know what I, as an old student of history and politics, and as one whose motto through life has been make allowance for everyone; am compelled to think after eighteen months of war. But I am still trying to make allowance, and your note helps thereto. Differ we must, but I feel the tenderest good will towards you... Difference of opinion is the salt of life.'

Paradoxically, the purely tactically aspect of battle held a distinct fascination for Fowler. As he himself put it: 'I had always been peacefully fond of warlike operations;' so much so that he devoted a chapter of *Kingham Old and New* to a detailed description of the extensive army manoeuvres which took place around Kingham on 15 November 1909. This interest persisted throughout the most harrowing months of 1915, as can be seen from the unpublished paper on the Battle of Waterloo, which included discussion on world war generally, and was written in June of that year.

Fowler's devotion to the village school, always intense, took on a practical form towards the end of 1915, at the same time providing him with a badly-needed outlet for his energies. After lamenting, in a letter to his friend P.E. Matheson, that he was too old and infirm to serve his country at such a critical time, he consoled himself: 'What I can do in that way I seize upon; e.g. as our schoolmaster has gone off to training, I am taking the school in hand, and have just been telling the older children how they too can help in the great work by keeping the school up to the mark.'

This practical involvement with the education of the Kingham children continued intermittently throughout the rest of Fowler's life. In *Essays in Brief for War-Time*, published by Blackwell in 1916, he recorded how he talked for half an hour to the older children about Shakespeare on the 300th anniversary of his death: 'I tried to make them feel that he really belonged to us, looked up at our hills from the Avon valley, spoke the same speech as our folk... He was... a man of the people; he came of a family of small farmers, like Mr Cook beyond the allotments yonder, and his father ran a variety of small trades, like our friend Mr. Eaton.' *Essays in Brief*, written during the agonising time of the battle of Verdun, brought some degree of relief to Fowler, both as a vehicle for articulating his thoughts on German scholarship and militarism and in providing the absorbing challenge of recording his views on a variety of matters dear to him as his life entered its seventieth year. The twenty short essays range in subject matter from the German pieces to his favourite game, backgammon, with diversions into ornithology, literature and religion as well as a tribute to *Two Ideal Schoolmasters*, F.E. Thompson of Marlborough and his best friend Sidney Irwin.

In his biography of Fowler, Dr. R.H. Coon gives a detailed account of daily life in the Fowler household at this time, as described to him by Emma Toon and Cordelia Holloway, both of whom had served Fowler and his sister in a domestic capacity for many years: 'He (Fowler) was called at seven-thirty with a cup of tea. Before breakfast, which was at eight-thirty, he went to church nearby for devotions. Breakfast finished, he read *The Times* and then set out, wearing his brown Shetland shawl, for a walk. He worked from ten till one, with instructions not to be disturbed under any circumstances. Though the hours were short he worked with intensity. Then came lunch and another and a longer walk, bringing him home in time for a nap before tea at four-fifteen. After the always important function of tea was finished, there was a short stroll (in the summer) and work until seven-thirty... He walked in bad weather, in the rain. When the weather was too severe his exercise was obtained by going up and down stairs. At times he varied the afternoon walks with golf... Always punctual himself, he expected the same of his household and visiting friends. Twenty minutes early he would be dressed for dinner and would sit down at the piano awaiting the call. The evening was devoted to music or to reading – often he read aloud for half an hour. When guests were not present, he would go out two or three times during the evening to observe the weather signs or to look at clouds or stars. And the day was concluded with a chapter or two of light reading in bed.

Spring cleaning bothered him. He was anything but neat and regarded it as unnecessary. His books would be dusted when he was absent; he would not know that it had been done. Or he would offer Cordelia ten shillings if she returned the books to their right places.'

On 8th January 1917, Alice Fowler died suddenly, leaving her brother, whose own health had by now begun a gradual yet permanent decline, benumbed with grief. Brother and sister had been close throughout their lives and had shared the house at Kingham since the death of their father in 1899. Fowler wrote to a friend: 'She was indeed a wonderfully beautiful soul, and I have never known this village so deeply moved as by the loss of her. Wherever her fragile form appeared, everyone knew that she was going on some errand of mercy. It is a great comfort to think that she passed, as she had lived, without any pain whatever, while she was dressing, after playing a whole quintet of Mozart with me the night before.'

Fortunately for Fowler, his two half-sisters, Emily and Florence, who came over to Kingham from their home at Malvern to support him during the weeks immediately following Alice's death, eventually decided to make their home with him. Fowler was beside himself with gratitude: 'Our relations have every day become more intimate, and I have learnt to appreciate the abilities of Emily, and the excellent management and good sense of Florence, as I never did before. Truly I am a fortunate man.' Emily was to survive him by twenty years, dying in 1941 at the age of eighty-eight; Florence, seven years her junior, lived in Kingham until her death, at the age of ninety-six, in 1956.

Although he had given up tutoring in 1908, Fowler had been persuaded to resume again during the war years on the understanding that his pupils would travel up to Kingham for their tutorials. By coincidence his last pupil, whom he tutored up until 1917, was H.M. Last, later Camden Professor of Ancient History at Oxford, who had never heard Fowler lecture yet who said of the old don's abilities as a tutor: 'He was very stimulating. His standard was so high that it made one want to do his best.'

The war continued to exert a powerful influence on Fowler's life and thinking throughout its duration. The loss of so many of his pupils and friends, as well as the growing toll of young village men, served to intensify his loathing for German military imperialism. He found a somewhat uncharacteristic vehicle for his beliefs in May 1918, in the form of a patriotic manifesto, intended to raise the morale of the village people, deadened after four years of conflict. In it, he described the struggle as one 'of Right against Might, for Liberty against Tyranny, for Good against Evil, for God against the Devil.' He went on to proclaim: 'We are fighting for the liberty and happiness of the people who are to come after us, not only where English is spoken, but over the whole world. So we are fighting for such a good cause as no people has ever yet had to fight for.'

Despite his declining health, the loss of his sister and his distress at the prolongation of, and suffering caused by the war, Fowler pursued his Virgilian studies with steadfast determination. He followed his *Gathering of the Clans* with *Aeneas at the Site of Rome*, which was published in 1917, and then turned his attention to the twelfth book of the *Aeneid*, which Blackwell published under the title of *The Death of Turnus* at the time of the cessation of hostilities at the end of 1918.

The year 1919 saw a great honour conferred upon Fowler, one which gave him immense pleasure, that of President of the Classical Association. Yet his letter of acceptance to his old friend Professor Conway reveals all his old diffidence and modesty: 'The honour is great – but alarming. I am reminded of Lummy (his dog) prodding a poor old snake out of his hole, where he (or rather she) was peacefully thinking of laying eggs... It is worth living another year for, but I am too old and deaf!'

He continued in this vein, making light of his infirmities, in another letter, dated January 1920, to Professor Gardner: 'Your faculties are as bright as ever. I wish mine were like them. I sometimes feel that my brain is not what it was. It won't supply me with ideas, poor thing. It is getting barren, it bears no fruit worth eating. It dislikes thinking, and wants me to read nice easy books, poetry, and literature. It says, you have driven me hard since 1900, and written at least eight books, the world needs no more of yours.'

Yet even in this, the penultimate year of his life, Fowler was absorbed in preparing his final book *Roman Essays and Interpolations* for the press and in writing his essays on Mozart for the publication *Music and Letters*. He also compiled a detailed paper on the history of Kingham parish church, which he read to a party of visiting students later in the year.

His *Reminiscences* and letters to friends occupied his thoughts and energies to within a few weeks of the end of his life. These letters ranged freely over a host of subjects, from world affairs to observations on bird life. He wrote to Professor Gardner in January 1921: 'I think it would be a good thing if we could all come to be less national in our outlook; and more international... Before the war I used to think that we Britons were ahead of other peoples in this respect, simply by reason of our Empire, and that the imperial state was an advance in political history. But the war knocked all these ideas about and developed in ourselves the old national feeling again. So the League of Nations has not really begun to interest people, though I look upon it as one of the most wonderful things of the day.'

Robert Bridges, the Poet Laureate, and a friend over many years, had asked Fowler to compile a list of the local names of British birds for the Society for Pure English, a task which he relished but lacked the strength to carry out. The love of birds remained with him to the end; in his last letter to his old friend A.H. Macpherson, dated 30th April 1921, he inquired about the feeding habits of the song thrush. He missed music keenly: 'My chief sorrow now is that I cannot hear the piano well enough to enjoy it.'

In February, his last writing for publication, an obituary of his friend Professor L.C. Miall, with whom he had worked on an edition of Gilbert White's *Selborne*, had appeared in *Nature*.

He concluded dictating his *Reminiscences* around the time of his seventy-fourth birthday on May 16th. By now his deteriorating health permitted only a short time out of doors, sitting in a wicker chair in his garden. He had to be carried upstairs on account of the weakness of his heart. Early in June his half-sisters noticed a decline in his mental faculties and from then onwards he only occasionally knew them.

He died on June 15th and was buried beside his sister Alice in a simply-inscribed grave near the porch entrance of the village church in which he had worshipped for nearly half a century.

CHAPTER 12

"Hence all is mild wisdom, humour, and a style that is perfect, both otherwise and because it is the man." (A friend of Warde Fowler)

A GOOD MAN REMEMBERED

'Warde Fowler, if perhaps not that rarest of beings, a great man, was certainly great as a man; and the news of his death will have meant to many, as it did to me, the loss of a friend.'

Julian Huxley's words, contained in a tribute to Fowler published in *British Birds* in November 1921, were typical of many expressions of appreciation which appeared in the press during the time immediately following his death. Huxley had been one of the many young scholars who had benefited from Fowler's habit of inviting like-minded men over to Kingham to share his home and companionship: 'We would tramp the meadows or the hills all the afternoon, and in the midst of bird-watching or stories of birds, he would break off to tell me the history of one of the fields, or to discuss the agricultural system of Kingham in feudal times. He was interested in the place and its history because he was unable to remain uninterested in any of the people or things with which he came into contact.'

The writer of his obituary in *The Times* paid tribute to: 'A rare and sensitive nature; he attached to himself all who came near enough to know him and his chosen pupils were his friends for life.'

The same writer assessed Fowler's achievements as tutor and historian: 'Warde Fowler's teaching and writing were marked by a combination of exact scholarship with a fresh and many-sided sympathy with life. It was a natural instinct with him to penetrate beneath the formal framework of history to the personality and interests not only of the chief figures but the ordinary men of the time. He loved to hand on the same alertness and humanity of outlook to his pupils and readers.'

In his obituary of Fowler in *The Observer*, John Sargeaunt paid particular attention to Fowler's success in combining his study of Virgil with the arduous task of teaching, one in which lesser men had failed: 'There was no such weakness in Fowler; and he was a learner to the last. ...Had all teachers the power and the freshness of Fowler, the study of the ancient classics would be in no danger.'

Fowler's lifelong affinity with the world of science was summarised by Professor E.B. Poulton, writing in *Nature*: 'Warde Fowler was one of the men we can least spare – a classical scholar of distinction and a writer of great charm who sympathised warmly with the aims and methods of science, and strove to give them a larger place in his University.'

Nor was recognition of Fowler's talents as a writer confined to his obituaries; thirteen years after his death, a reviewer of Professor Coon's biography of Fowler in *The Times Literary Supplement* described him as: 'a charming writer, with a completely unaffected, lucid, musical, personal style.'

Of all these testimonials to Fowler's many and varied attributes, none gets closer to the essence of his character than that which appeared in the *Oxford Magazine*, the journal whose pages his work had so often graced, on 20th October 1921: 'By his friends it is the man that will be first remembered, his delightful personality, and the sincerity that marked all his words and actions. In his leisure hours no one was ever more full of fun, or told a better story.' The anonymous writer clearly knew Fowler well: 'Like all strong characters he had his dislikes, and he never tolerated men who had no <u>stuff</u> in them.'

Assessing Fowler's role in the Kingham village community, he wrote: 'Many villagers will remember the sympathy with distress, and his readiness to relieve genuine want, and to help genuine aspirants to a higher sphere. Naturally they were incapable of appreciating the finer shades of his character or of recognising that they had a man of rare genius and rare sympathies in their midst.'

Sadly, almost all such memories and impressions of Fowler by the village folk have gone unrecorded and are lost forever. The few remaining, recalled by the dwindling band of those who could still remember him, serve at least to confirm those qualities of compassion, generosity and endearing eccentricity for which he was noted in the wider circles in which he also moved.

The Phillips family in particular, cherished his memory with respect and affection. The seven children of George Phillips, yeoman famer, dissenter, parish clerk and by popular acclaim the Grand Old Man of Kingham, they all won scholarships to grammar school and the five sons went on to Oxford before entering the teaching profession. Ronald Phillips recalled that: 'W.W.F's testimonial helped get me into St. Edmund Hall. Our vice Principal talked to me about him and obviously had great respect and admiration for him.'

Norman Phillips remembered as a schoolboy writing to Fowler, who was away lecturing in Edinburgh at the time of the violent thunderstorm described in *Kingham Old and New*. Later, after entering Burford Grammar School, he recalled vividly an incident illustrating Fowler's characteristic generosity: 'My father had a small field which adjoined Mr Fowler's house and garden. I was making my way up this field, amusing myself the while by hitting a small chunk of wood with a broken branch from a nearby tree. I think I had given myself six hits to reach the field gate. Suddenly I heard a clear voice behind me enquiring if I was playing golf. I turned and saw Mr Fowler looking at me through the hedge. I admitted that it was a bit like golf, to which he said 'Come round to my house and I will give you some clubs.' I went round at once and he gave me two clubs, a brassie and an iron, and several balls. He thought I should enjoy the game better with better tools. My brothers and I made a two-hole golf course in the close and enjoyed many hours of glorious, if amateurish, golf.'

May Phillips, who for many years contributed a weekly column headed *In Country Ways* to the *Oxford Times*, described Fowler as: 'this gentle, kindly man, who would, if he caught her up in the street, carry his washerwoman's basket of laundry to her destination, and was always ready to listen to children's excited talk of the birds and flowers they had seen.'

Alfred Jarvis remembered Fowler's influence in the field of natural history: 'I was a scholar at Kingham before the First World War and often met Warde Fowler on the Kingham Hill road and along the Marsh. I have been with him searching a hedge or dropping stones into a pond and counting the resulting circles. He was a great naturalist and full of knowledge of birds, especially the warblers. I think he knew all the bridle paths for miles around and must have walked many miles in the district. He had a great love of Kingham and its people. He was a gentle, kindly man who possessed a great sense of humour.'

Sir Basil Blackwell, in recalling that *A Year with the Birds* was one of his father's first publications and that *Kingham Old and New* was the first book he himself published, said of Fowler: 'I knew Warde Fowler and loved him well... I had very happy relations with this dear good man until his death.'

In addition to such tributes contained in obituaries and the fragmented reminiscences of those who knew him, the name of William Warde Fowler is perpetuated in numerous books dealing with the subjects dear to his heart. A good number of these references date from his lifetime; others such as Earl Grey of Fallodon's tribute in *The Charm of Birds*, already quoted, were written shortly after his death.

Fowler's old Marlborough friend, H.A. Evans, paid him an anonymous yet unmistakable tribute in his *Highways and Byways of Oxford and the Cotswolds* in 1905: 'the joys of Kingham living have, however, been the theme of a subtler pen than mine, and to the lover of birds and books I need say no more.' Three years earlier, W.L. Mellersh in his introduction to *A Treatise on the Birds of Gloucestershire* stated that: 'The writer has tried to follow the example set by a valued friend, Mr W. Warde Fowler, keeping to the living habits of birds as affecting their geographical and general relationship to the county.'

Fowler's aptitude at capturing the idiosyncrasies of individual species of birds was acknowledged by many of his contemporaries. One such admirer, Eric Parker, himself a writer of some stature on rural affairs, referred in his *English Wild Life* (1929) to the chaffinch's 'happy little roulade of mating time, which Warde Fowler, kindliest of Oxford dons, used to say reminded him of a bowler's run up to the wicket.'

In a chapter headed *Nineteenth Century Enclosures* in *A Corner of the Cotswolds*, published in 1914, the historian M. Sturge Gretton, referring to Fowler and the recently-published *Kingham Old and New*, stated: 'How ripe and informed that authority is, everyone interested in rural England is privileged now to judge for himself.' Similar sentiments were expressed by H. L. Trollope in her book *Old Days in Country Places* (1923): 'Dr Warde Fowler's *Kingham* shows what interest and what history one little district can provide and that its study is the work of a happy lifetime.'

Lying as it does off the main tourist routes across the Cotswolds, Kingham tends to have been overlooked by most of the topographical writers of the region. One notable exception is the volume on Oxfordshire in Arthur Mee's *The King's England* series, in which Fowler is aptly remembered: 'William Warde Fowler gave Kingham an abiding place in literature. An Oxford don, a great authority on Roman history and customs, he made this place his country home and the holiday home of his students for nearly half a century. Here, in spite of defective sight and hearing, he studied bird life with the ardour and perception of a Gilbert White, and wrote enchantingly of it. His bird books, like his book *Kingham Old and New* are classics.'

In his *Thirteen Rivers to the Thames* (Dent, 1964), Brian Waters wrote: 'The charm of the Oxfordshire village is revealed in the natural 'tempo' of a book, written by a scholar who lived there for 30 years before he wrote *Kingham Old and New*. William Warde Fowler (like Gilbert White) was an Oxford don and a bachelor, whose affections were absorbed in ornithology.'

It was appropriate that when one of the young men he had be-friended at Lincoln College himself wrote a book on birds, the work should be dedicated to Fowler. The author was Walter Garstang, Professor of Zoology at the University of Leeds, whose book *Songs of the Birds* was published shortly after Fowler's death. Garstang chose to honour Fowler's memory in verse:

'Three things, old friend, in youth revealed,
With you are interwoven –
A common room, rich walks afield,
Rare evenings with Beethoven.

They came as joys at twenty five,
When life was far from thrifty;
They prove delightful still, and thrive
In memory at fifty.

To you, then, who in those far days
Could overlook my blindness,
I dedicate these rustic lays –
Late blossoms of your kindness.'

In a more recent ornithological study, *The Birds of Berkshire and Oxfordshire* (1966), Dr. M.C. Radford said that Fowler was: 'One of the first to aim at observing rather than obtaining speci-mens. He left books, *A Year with the Birds* and *Kingham Old and New* which still remain delightful reading. Perhaps he should be best remembered for his work on the Marsh Warblers at Kingham. Aplin said he was indebted to him for much help and information and certainly Warde Fowler inspired many others.' Fowler's part in encouraging the 'new approach' to bird study – watching rather than shooting – was also acknowledged by David Elliston Allen in his book *The Naturalist in Britain* (Allen Lane, 1976).

'Few fellows,' wrote Dr V.H.H. Green in *The Commonwealth of Lincoln College*, 'excited so strong a pull on the affection of undergraduates or were so esteemed by their colleagues.' Dr. Green went on to quote the late Professor Rose, who dubbed Fowler as 'the best type of Victorian don, a quiet and unassuming gentleman, English to the core; a lover of the country... deeply religious, and totally strange to fanaticism.'

Little wonder that on hearing of his death, one of the many who loved him wrote: 'To one of his friends at least, and the same is doubtless true of others, it seems as if a light has gone out of their lives.'

CHAPTER 13

One great glory of Kingham is in the footpaths that lead in
every direction, which are not merely short cuts from one point
to another, but often stretch away over grass-fields for miles,
without once bringing you to the sight of a road.

THE FOWLER COUNTRY TODAY

A good many changes have taken place in and around the village of Kingham since Warde Fowler's time. Nevertheless, the discerning visitor will find a number of associations bridging the years since his death.

The plain grave by the church porch bears the inscription:

Alice Augusta Elizabeth Fowler
Born Dec 26 1848
At Rest Jan 8 1917

William Warde Fowler
Born May 16 1847
At Rest June 15 1921

A brass plaque on the wall of the nave is engraved:

'To the memory of William Warde Fowler, Fellow of Lincoln College, Oxford. Born 16th May 1847, died 15th June 1921 & of his sister, Alice Augusta Fowler, born 26th December 1848, died 8th January 1917. Both for many years resident in this parish. Dedicated in affectionate remembrance by their nephews and nieces'.

The stately former rectory adjacent to the church, built by William Dowdeswell the Elder in 1688 and home of Fowler's friend Rev. Samuel Davis Lockwood, is now a private residence.

Immediately across the road stands the Corner House, where Fowler lived during the early years of his residence in the village.

Fowler House itself has changed little externally since it was built in 1879. In his will, Fowler left the house to Lincoln College, his wish being that after his sisters' deaths it should become a weekend home of rest for the fellows and a hostel of study during vacations. Sadly, this wish was never realised; the house was sold and has since been a private residence.

Within recent years, however, Fowler's name has been perpetuated in the village by the naming of a street of new houses adjacent to his old home after him.

Across Church Street stands the delightful Old Rectory Cottage, once the home of Fowler's friend Colonel John Barrow and known in his days as the 'Castle' or 'Monte Rosa'. Later, during the war years, Lady Mary Kidd, daughter of Lord Lothian, lived there. She became friendly with Fowler and his sister. He dedicated his *Essays in Brief for Wartime* to 'A Good Friend and a Good Listener, Lady Mary Kidd.' Rectory Cottage eventually became the home of the radio personality Frederick Grisewood, the son of the Rector of the neighbouring village of Daylesford. It was Grisewood, in association with Basil de Selincourt and his wife, the novelist Anne Douglas Sedgwick, who established the tradition of dramatic productions performed in the village hall erected by Charles Baring Young, squire of Daylesford and founder of Kingham Hill Homes (now Kingham Hill School).

The council school, in which Fowler took such an active interest following its opening in 1912, is now a county primary school, catering for the needs of the children up to the age of eleven from

Kingham and several surrounding villages. It was among the first in Oxfordshire to establish a nature conservation and study reserve. This is situated on an area of ground adjacent to the school in what was formerly part of the village allotments. Here children planted trees, made and marked paths and carried out natural history projects, thus maintaining a lively involvement in field studies in the truest tradition of William Warde Fowler.

Despite the loss of considerable quantities of woodland, hedgerows and wetland habitat brought about by changes in land use and management, the Kingham region is comparatively rich in areas specially designated for wild life conservation. In addition to the small village school reserve, another area of woodland has been established along the course of the disused railway branch line near Kingham Station, and the Berks, Bucks and Oxon Wildlife Trust manages an extensive nature reserve known as Foxholes, between Kingham and Idbury, consisting of 160 acres of woodland and meadow alongside the River Evenlode and traversed on its eastern side by the long distance footpath, the Oxfordshire Way.

On the subject of footpaths, it is worth noting that several of those regularly followed by Warde Fowler are still clearly definable and the walker following them can assess the current distribution of wildlife as compared with that he recorded a century or so ago.

Perhaps the most productive of these walks in terms of species, both during Fowler's lifetime and at the present day, is that over Coxmoor, which receives an entire chapter in *Kingham Old and New*. This walk commences as a wide trackway at the opposite end of the playing field (Bury Pen Close) from the church and passes the foot of the gardens of a housing estate (built after Fowler's death). Soon the houses are left behind and after crossing a stile, the walker enters open fields with splendid distant views to the right of Oddington Ashes, an extensive tract of mixed woodland, and Stow-on-the-Wold. Ahead rises Icomb Hill surmounted, in Fowler's day by an observation tower, but now by a television mast. Away to the left stretch Bruern Woods, of which the Foxholes Nature Reserve forms part.

After passing through an area of hawthorn scrub, rich in flowering plants and butterflies on summer days, the walker reaches the Paddington-Hereford railway line, north of Kingham Station. Known as Chipping Norton Junction in Fowler's day, this station is now merely a halt between Charlbury and Moreton-in-Marsh. Gone are the distinctive architecture and ornate wrought iron of the heyday of the Great Western Railway; but the traveller can still, like Fowler, clatter along down to Oxford in half an hour through delightful countryside with memorable glimpses of the winding Evenlode, deep woods, leafy lanes and half-hidden Cotswold-stone cottages.

Gone too, a couple of miles or so up the line and over the county boundary, is the little station of Adlestrop, where Edward Thomas, sitting in a compartment of an express train that halted unexpectedly in this tranquil landscape, heard and immortalised the birds of Oxfordshire and Gloucestershire.

Soon after crossing the line, the footpath enters a large meadow via a stile and skirts a low-lying tract of marshy woodland, enclosed by a barbed-wire fence. This was formerly the osier bed in which Fowler recorded the nesting of the rare marsh warbler between 1893 and 1904. A few ancient osiers still survive, alongside a scattering of sallows, ashes and hawthorn. The river Evenlode now comes into view and is spanned by a wooden footbridge. It is a sluggish reedy little river, its waters muddied by the sediments of a valley far removed in composition and character from the true Cotswold rivers – Windrush, Dikler, Coln, Leach – that feed into the Thames from the limestone uplands.

This is the celebrated meadow about which Fowler wrote: 'Of the many interesting things I have seen in Coxmoor, things that send you home with curiosity awakened and a new sense of life. No other spot in all our region has given me so many pleasant surprises as this.'

Crossing the footbridge, the walker enters Gloucestershire and follows the footpath over the embankment of the disused Cheltenham railway line, to emerge across a low-lying water meadow at the village of Bledington. A number of walks of varying lengths are possible from here; alternatively the return to Kingham can be made along the road passing the railway station.

Second only to Coxmoor, the Yantle, a large meadow lying between Swailsford Brook and the now disused branch line to Chipping Norton, was Fowler's happiest hunting ground. It can best be reached by taking the Churchill - Chipping Norton road eastwards out of the village beyond the green and school. Apart from a footpath to Churchill at the foot of the hill, however, no right of way exists across the Yantle.

Another, longer ramble which affords pleasant, easy walking along bridleways and minor roads can be taken along the track to Swailsford Bridge, near Cornwell and back to Kingham via Whitequarry Hill, a favourite botanising location of Fowler and his sister and also the site of the quarry from which the stone for the building of Fowler House was obtained. Kingham Field, a sixty-acre expanse on the left of the road descending from below Kingham Hill School to Kingham village, has yielded substantial quantities of bronze ornaments, Roman coins and pottery over the years, much of which was excavated by Mrs Grisewood, wife of the rector of Daylesford and mother of Freddie, during the first decade of the last century.

A more rigorous and challenging walk, for which boots or stout shoes are advised, involves taking the straight track leading to the right from West End, at the bottom of West Street, towards Daylesford. This track can be followed either directly to Daylesford or by taking a left fork near a group of farm buildings; an alternative route can be taken over the railway and via Bledington Heath and Oddington Ashes. The ground along this stretch is often waterlogged even in summer, but the sense of remoteness is strikingly impressive. Daylesford was part of Worcestershire during Fowler's lifetime. The churchyard contains the grave of the statesman Warren Hastings. The tree-lined hedgerow marking the county boundary between Gloucestershire and Oxfordshire is worth noticing on the left of the road back to Kingham, as it contains examples of many different species of trees and shrubs.

Finally, a short walk in Churchill parish, two miles or so from Kingham, along a lane known to Fowler as 'Butterfly Lane', though marked 'Besbury Lane' on the large-scale Ordnance Survey maps. This lane regularly yielded him a chalkhill blue butterfly, and as he recorded in *Kingham Old and New*: 'All the common kinds which luckily include some of the most beautiful, rejoice in the upper part of this lane.'

The walk can be followed by leaving Churchill on the Chipping Norton road and parking on the verge at the entrance to the lane, which leads off to the right by a wood, and is marked 'Conduit Farm and Stud.' The lane is metalled at first but soon reverts to a rougher surface and climbs steadily between ancient hedges passing a tumulus mound on the right before falling away and eventually meeting the Burford – Chipping Norton road. As in Fowler's day, the summer roadside herbage provides excellent sustenance for butterflies: knapweed, scabious, bindweed and ladies bedstraw flourish.

Retracing one's steps, with only a singing skylark for company and stone walls replacing hedges over the slopes towards Chipping Norton, a splendid view of the Cotswold escarpment appears ahead. Stow church tower rises above the wolds to the north, with the communications mast on Icomb Hill providing another landmark slightly southwards. The former Central Flying School, Little Rissington, now the site of a new village, Upper Rissington, lies directly ahead, crowning the ridge between the Windrush and Evenlode valleys, and the sugar-loaf wooded knoll known as Blackheath Clump commands the skyline to the south. The greater part of Warde Fowler's natural history writings owe their origins to the countryside lying within this view.

We know from his writings and especially from his entries in Colonel Barrow's celebrated log books, that in his younger days Fowler undertook many more walks around his beloved Kingham, some of them involving arduous tramping over inhospitable terrain. A fair amount of this walking must have been along roads, dusty in summer, muddy in winter, which in Fowler's day were comparatively free of traffic; nevertheless achieving such destinations as Adlestrop Hill, the Rollright Stones, the Merrymouth Inn and Sarsden Pillars demanded both determination and physical stamina, and Warde Fowler, for all his slightness of build and indifferent health, must have possessed ample reserves of both these qualities.

That he knew this region of our Midlands landscape with a rare and sensitive understanding is beyond question. In following his footsteps, we too can hope, even after a century of change, to gain an insight into its distinctive, almost indefinable charm.

CHAPTER 14

*"Only those could be admitted within the magic circle of his
closest friends who knew the spells of wild life." (A friend of
Warde Fowler)*

WILDLIFE – A CENTURY OF CHANGE

'As I read once more my ancient book *A Year with the Birds*,'
Fowler wrote in *Kingham Old and New*, 'I find that much of what
is said there about the birds of Kingham does not strictly hold
good of the Kingham of the twentieth century.'

Bearing in mind that the changes to which Fowler refers took
place over a period of less than thirty years, prior to the great wa-
tershed of the First World War, it would hardly be surprising considering the profound nature of the revolution in farming methods
and land use generally, to find that the state of the flora and fauna
of the Kingham district had undergone an almost unrecognisable
change since his words were written.

Happily, however, although the status and distribution of certain
animals, birds and plants have declined significantly since
Fowler's time, the vast majority are still present, albeit in reduced
numbers and the gaps left by the loss of certain species have been
filled, not always satisfactorily from a human standpoint, by certain opportunist newcomers.

The most significant change in the Kingham landscape is undoubtedly the depletion of woodland. One has only to compare
views of the village and its surrounding countryside as depicted

in photographs taken in the early decades of the twentieth century to realise the extent of the impact resulting from the widespread felling of woods, clumps and individual trees. True, this lamentable trend has been halted, reversed to some degree even, during the past twenty or so years; nevertheless the clearance of such extensive tracts as Churchill Heath, the demise of the elms and the loss of countless other trees across the locality can only have had a detrimental effect on the status of wild life generally.

The hedges, by contrast, appear to have undergone little change since Fowler's time. A few have been grubbed up to extend the arable fields, especially in the Kingham Hill area, but bearing in mind that the enclosure of the open fields did not take place until 1850, the hedge chequerwork that Fowler knew remains essentially unchanged today, except that the ancient craft of hand-laying tends to have been superseded by mechanical cutting.

It is in the fields themselves that the full impact of the modern agricultural revolution has made its presence felt. No corner of waste or water-logged ground is too small for clearance or drainage; none but the most persistent of the cornfield flowers can resist the onslaught of the weed killer. The harvest field is stripped of its bounty by a swift, round-the-clock efficiency that the gleaners – man and animal – once completed in timeless late-summer days.

It seems appropriate to begin a brief survey of the wildlife of the Kingham area at the present day by considering the birds, and especially the small birds of which Fowler was so passionately fond. There is no record of the marsh warbler recolonising the osier beds near the river Evenlode. The sedge warbler, however, can be heard chattering during both day and night from the damp margins along the valley and both chiffchaff and willow warbler, together with the blackcap, whitethroat, lesser whitethroat and occasionally the grasshopper warbler return each spring to add a rich diversity, both in song and plumage, to the bird life of the area.

Sadly, two other of Fowler's favourite summer visitors, the redstart and the whinchat, are rarely recorded in the vicinity of Kingham now, a decline which Fowler himself observed during his residence in the village. The same is true of the yellow wagtail, now, like the wheatear, a scarce passage migrant. A singing nightingale is reported in the locality occasionally in spring and the tree pipit and spotted flycatcher breed in suitable locations each year.

Of the resident song birds, the skylark, blackbird, song and mistle thrush, robin, wren, dunnock, chaffinch, greenfinch and yellowhammer all maintain fair numbers. Goldfinches and linnets are fairly abundant and the bullfinch, corn and reed buntings have been recorded in recent times. The various members of the tit tribe manage to maintain a stable population, together with green and greater spotted woodpecker, treecreeper and goldcrest. Nuthatches, however, are scarce. Grey wagtails appear from time to time along the Evenlode and its tributary streams and kingfishers are still seen occasionally.

Changes in land use and the increasing efficiency of agricultural techniques have hastened the decline, disappearance even, of several species of birds, among them the red-backed shrike, the stonechat and the corncrake. By contrast, two introduced species, the red-legged partridge and the little owl, have become successfully established in the area since Fowler's time. Four species of gull: the black-headed, common, herring and the lesser black-backed, have also undergone a dramatic increase in numbers and can be seen scavenging on tips and following the plough; while the collared dove, unknown in Britain in Fowler's day, is now a ubiquitous and in the view of many, an excessively vocal, resident around Kingham and its neighbouring villages.

Apart from the cornfield flowering plants alluded to earlier, most of the plants referred to by Fowler can still be found today, together with the range of butterflies with which they are associated. Without doubt, however, the extent of suitable habitat – heath, quarry, pasture, verge, woodland – has shrunk over the years and continues to do so, thus rendering the task of locating them increasingly difficult.

What of the future? This little – and little-known – corner of England is hardly likely to win acclaim from many but a few discerning admirers, who either know it through being fortunate enough to live within its limits, or who discovered it by chance, or who were led to it by the writings of William Warde Fowler. On one thing for certain, these happy few will agree – that this fragment of midland England exerts a hold on the senses that is so compelling as to be almost irresistible.

BIBLIOGRAPHY

Barrow, John — *Log Books 1869-1882* (Bodleian Library, Oxford)

Coon, Raymond H. — *William Warde Fowler. An Oxford Humanist* (1934)

de Selincourt, Basil — *Anne Douglas Sedgwick. A Portrait* (1934)

Evans, H.A. — *Highways & Byways in Oxford & the Cotswolds* (1908)

Fowler, W. Warde — *A Year with the Birds* (1886)

Fowler, W. Warde — *Tales of the Birds* (1888)

Fowler, W. Warde — *Summer Studies of Birds & Books* (1896)

Fowler, W. Warde — *More Tales of the Birds* (1902)

Fowler, W. Warde *In Praise of Rain* (included in *The Book of the Open Air*, ed. Edward Thomas, 1907)

Fowler, W. Warde — *Kingham Old and New: Studies in a Rural Parish* (1913)

Fowler, W. Warde — *Essays in Brief for War-Time* (1916)

Fowler, W. Warde — *Reminiscences* (1921)

Fowler, W. Warde — Numerous articles in the *Oxford Magazine* (1884-1908)

Garstang, Walter — *Songs of the Birds* (1922)

Green, David — *Country Neighbours* (1948)

Green, V.H.H. — *The Commonwealth of Lincoln College* (1977)

Gunther, R.T.	*The Oxford Country* (1912)
Grey, Earl of Fallodon	*The Charm of Birds* (1927)
Hudson, W.H.	*Birds and Men* (1901)
Hudson, W.H.	*Adventures among Birds* (1913)
Jarvis, Alfred	*Fifty Years of Kingham Hill* (1936)
Lainchbury, Ernest	*Kingham the Beloved Place* (1957)
Lockwood, E.D.	*The Early Days of Marlborough College* (1893)
Mee, Arthur (Ed)	*Oxfordshire* (King's England Series, 1939)
Mellersh, W.L.	*A Treatise on the Birds of Gloucestershire* (1902)
Ottewell, Gordon (Ed)	*Warde Fowler's Countryside* (1985)
Ottewell, Gordon	*Literary Strolls in the Cotswolds & Forest of Dean* (2000)
Ottewell, Gordon	*The Evenlode. An Exploration of a Cotswold River* (2004)
Parker, Eric	*English Wildlife* (1929)
Radford, M.A.	*The Birds of Berkshire & Oxfordshire* (1966)
Sturge Gretton, M.	*A Corner of the Cotswolds* (1914)
Trollope, H.L.	*Old Days in Country Places* (1923)
Waters, Brian	*Thirteen Rivers to the Thames* (1964)

INDEX

A

Ainsworth, Harrison *12*
Allen, D.E. *113*
Andersen, Hans *8*
Anderegg, Johann *25, 43, 55*
Acland, Arthur *38*
Aplin, Oliver *61, 65, 67*
Arch, Joseph *49*
Arnold, Matthew *83*
Austen, Jane *11, 46, 97*

B

Bacon, John *9*
Bacon, Elizabeth *10,13,15,23,25*
Barrow, John *27, 28, 33, 39, 41,
91, 115, 120*
Baring Young, Charles *98*
Barnes, William *68*
Baynes, F.H. *25*
Beethoven, Ludwig von *31*
Blackwell, Sir Basil *110*
Brahms, Johannes *85*
Brick, Robert *84*
Bridges, Robert *104*

C

Cook, Keeper *63*
Coon, Raymond H. *32, 91, 100,
107*
Conway, R.S. *82, 103*
Crawley, Charles *20*

D

Darwin, Charles *97*
de Selincourt, Basil *85, 115*
Dickens, Charles *97*
Dill, Samuel *82*
Dowdeswell, William *114*

E

Elliott, Gilbert *56*
Eliot, George *97*
Evans, Herbert A. *18, 22, 110*

F

Farjeon, Eleanor *92*
Fatro, Professor *27*
Fowler, Alice *7, 13, 43, 52, 79, 95,
101, 105, 114*
Fowler, Emily *7, 43, 79, 101*
Fowler, Florence *7, 43, 79, 101*
Fowler, John *7, 26, 43*
Fowler, John C. (senior) *7, 9, 13*
Fowler, Thomas *21, 33, 46, 80*

G

Gardner, Percy *103*
Garstang, Walter *112*
Grey, Earl, of Fallodon *56, 110*
Green, David *88*
Green, V.H.H. *49, 82, 113*
Grisewood, Frederick *119*
Grisewood, Mrs. *115*
Gunther, R.T. *90*

H

Hammond, John *49, 50*
Hastings, Warren *119*
Henderson, B.W. *82*
Holloway, Cordelia *100, 101*
Housman, A.E. *98*
Hudson, W.H. *64, 65*
Huxley, Julian *106*

I

Irwin, Sydney *84*

J

Jarvis, Alfred *109*

K

Kidd, Lady Mary *115*
Kilvert, 'Gaffer' *11*, *13*

L

Lainchbury, Ernest *85*
Last, H.M. *102*
Lockwood, E.D. *84*
Lockwood, Rev. S.D. *42*, *88*, *114*
Lucas, E.V. *94*

M

Macpherson, Arthur *61*, *66*, *67*, *104*
Mann, H.E. *49*, *80*, *83*, *85*, *92*
Matheson, P.E. *99*
Mellersh, W.L. *110*
Miall, L.C. *61*, *104*
Mozart, Wolfgang A. *13*, *39*, *97*, *101*, *103*
Munro, J.A.R. *51*

N

Nettleship, Henry *21*

O

Ostler, R.S. *69*

P

Parker, Eric *110*
Pattison, Mark *21*, *33*, *47*
Pierce, Miss *8*
Phillips, George *98*, *108*
Phillips, May *109*
Phillips, Norman *109*
Phillips, Ronald *108*
Playne, Herbert *61*, *67*
Porter, Lucy *91*
Poulton, E.B. *107*

R

Radford, M.C. *112*
Rose, H.J. *82*, *113*

S

Sargeaunt, John *107*
Schubert, Franz *31*
Scott, Walter *97*
Sedgwick, A.D. *85*, *115*
Shakespeare, W. *97*, *99*
Sidgwick, N.V. *93*
Stewart, J.A. *30*
Sturge Gretton, M. *111*

T

Thomas, Edward *57*, *92*, *117*
Thompson, F.E. *18*, *100*
Toon family *41*, *43*
Trollope, H.L. *111*

W

Wallace, Alfred *97*
Waters, Brian *111*
White, Gilbert *35*, *61*, *63*, *88*, *91*, *93*, *94*, *104*, *111*

Also by Gordon Ottewell

Warde Fowler's Countryside (Edited)
The Countryside our Classroom
Square Peg: Memoirs of a Misfit Miner
Gloucestershire Countryside
A Cotswold Country Diary
Discovering Cotswold Villages
The Evenlode: An exploration of a Cotswold river
Theme Walks in Gloucestershire
Literary Strolls in the Cotswolds & Forest of Dean
Literary Strolls in Wiltshire & Somerset
Family Walks in the Cotswolds
Family Walks around Stratford & Banbury
Family Walks in South Gloucestershire
Family Walks in Northamptonshire
Family Walks in South Derbyshire

For children

Journey from Darkness
Tangleton